Cryptography and Its Applications in Information Security

Cryptography and Its Applications in Information Security

Editors

Safwan El Assad
René Lozi
William Puech

MDPI • Basel • Beijing • Wuhan • Barcelona • Belgrade • Manchester • Tokyo • Cluj • Tianjin

Editors
Safwan El Assad
Université de
Nantes-Polytech
France

René Lozi
Côte d'Azur University
France

William Puech
AUniversity of Montpellier
France

Editorial Office
MDPI
St. Alban-Anlage 66
4052 Basel, Switzerland

This is a reprint of articles from the Special Issue published online in the open access journal *Applied Sciences* (ISSN 2076-3417) (available at: https://www.mdpi.com/journal/applsci/special_issues/Cryptography_Applications_Information_Security).

For citation purposes, cite each article independently as indicated on the article page online and as indicated below:

LastName, A.A.; LastName, B.B.; LastName, C.C. Article Title. *Journal Name* **Year**, *Volume Number*, Page Range.

ISBN 978-3-0365-3767-2 (Hbk)
ISBN 978-3-0365-3768-9 (PDF)

Cover image courtesy of René Lozi

© 2022 by the authors. Articles in this book are Open Access and distributed under the Creative Commons Attribution (CC BY) license, which allows users to download, copy and build upon published articles, as long as the author and publisher are properly credited, which ensures maximum dissemination and a wider impact of our publications.

The book as a whole is distributed by MDPI under the terms and conditions of the Creative Commons license CC BY-NC-ND.

Contents

About the Editors vii

Safwan El Assad, René Lozi and William Puech
Special Issue on Cryptography and Its Applications in Information Security
Reprinted from: *Appl. Sci.* **2022**, *12*, 2588, doi:10.3390/ app12052588 1

Jaehyoung Park and Hyuk Lim
Privacy-Preserving Federated Learning Using Homomorphic Encryption
Reprinted from: *Appl. Sci.* **2022**, *12*, 734, doi:10.3390/ app12020734 5

Gianmarco Baldini, Jose Luis Hernandez Ramos and Irene Amerini
Intrusion Detection Based on Gray-Level Co-Occurrence Matrix and 2D Dispersion Entropy
Reprinted from: *Appl. Sci.* **2021**, *11*, 5567, doi:10.3390/app11125567 23

Satoshi Iriyama, Koki Jimbo and Massimo Regoli
New Subclass Framework and Concrete Examples of Strongly Asymmetric Public
Key Agreement
Reprinted from: *Appl. Sci.* **2021**, *11*, 5540, doi:10.3390/app11125540 47

Fethi Dridi, Safwan El Assad, Wajih El Hadj Youssef, Mohsen Machout and René Lozi
The Design and FPGA-Based Implementation of a Stream Cipher Based on a Secure Chaotic
Generator
Reprinted from: *Appl. Sci.* **2021**, *11*, 625, doi:10.3390/app11020625 79

**Evaristo José Madarro Capó, Carlos Miguel Legón Pérez, Omar Rojas,
Guillermo Sosa-Gómez and Raisa Socorro Llanes**
Bit Independence Criterion Extended to Stream Ciphers
Reprinted from: *Appl. Sci.* **2020**, *10*, 7668, doi:10.3390/app10217668 99

Rong Huang, Fang Han, Xiaojuan Liao, Zhijie Wang and Aihua Dong
A Novel Intermittent Jumping Coupled Map Lattice Based on Multiple Chaotic Maps
Reprinted from: *Appl. Sci.* **2021**, *11*, 3797, doi:10.3390/app11093797 119

Ricard Borges and Francesc Sebé
A Digital Cash Paradigm with Valued and No-Valued e-Coins
Reprinted from: *Appl. Sci.* **2021**, *11*, 9892, doi:10.3390/app11219892 141

Yun-Ciao Wang, Chin-Ling Chen and Yong-Yuan Deng
Authorization Mechanism Based on Blockchain Technology for Protecting Museum-Digital
Property Rights
Reprinted from: *Appl. Sci.* **2021**, *11*, 1085, doi:10.3390/app11031085 155

About the Editors

Safwan El Assad is Associate Professor (HDR) at Nantes University/Polytech Nantes, IETR (Institut d'Electronique et des Technologies du numéRique) laboratory, UMR CNRS 6164, VAADER team, France. From 1988 to 2005, his research activities concerned radar imagery and digital communications. With a background in radar imagery and digital communications, today his research largely focuses on chaos-based cryptography: Block/Stream ciphers; Keyed Hash Functions; Authenticated Encryption; Steganography and Watermarking systems. Awards (6): PEDR (Award for Doctoral Supervision and Research): (1994-1997; 1998-2001), PES (Award for Scientific Excellence): (2010-2013; 2014-2017), PEDR (2018- 2021; 2021-2025). https://scholar.google.com/citations?user=69Jk1jQAAAAJ&hl=fr.

René Lozi is Emeritus Professor at University Côte d'Azur, Dieudonné Center of Mathematics, France. He completed the PhD degree with his French State Thesis (on chaotic dynamical systems) under the supervision of Prof. René Thom (Fields medalist) in 1983. In 1991, he became Full Professor at University of Nice and IUFM (Institute for teacher trainees). He has served as Director of this institute (2001-2006) and as Vice-Chairman of the French Board of Directors of IUFM (2004-2006). He is member of several editorial boards of international journals. In 1977, he discovered a particular mapping of the plane having a strange attractor (now, commonly known as "Lozi map"). Nowadays, his research areas include complexity and emergence theory, dynamical systems, bifurcations, control of chaos, cryptography based on chaos, and recently memristors (physical devices for neurocomputing). He is working in those fields with renowned researchers from many countries. He received the Dr. Zakir Husain Award 2012 from the Indian Society of Industrial and Applied Mathematics during the 12th biannual conference of ISIAM at the University of Punjab, Patialia, January 2015.

William Puech received the diploma of Electrical Engineering from the Univ. Montpellier, France (1991) and a Ph.D. Degree in Signal-Image-Speech from the Polytechnic National Institute of Grenoble, France (1997) with research activities in image processing and computer vision. He served as a Visiting Research Associate to the University of Thessaloniki, Greece. From 1997 to 2008, he has been an Associate Professor at the Univ. Montpellier, France. Since 2009, he is a full Professor in image processing at the Univ. Montpellier, France. His current interests are in the areas of image forensics and security for safe transfer, storage and visualization by combining data hiding, compression, cryptography and machine learning. He is head of the ICAR team (Image and Interaction) in the LIRMM and has published more than 45 journal papers and 140 conference papers and is associate editor for 5 journals (JASP, SPIC, SP, JVCIR and IEEE TDSC) in the areas of image forensics and security. Since 2017 he is the general chair of the IEEE Signal Processing French Chapter. He was a member of the IEEE Information Forensics and Security TC between 2018 and 2020 and since 2021 he is member of the IEEE Image, Video and Multidimensional Signal Processing TC.

Editorial

Special Issue on Cryptography and Its Applications in Information Security

Safwan El Assad [1,*], René Lozi [2] and William Puech [3]

1. Institut d'Electronique et des Technologies du NuméRique (IETR), UMR CNRS 6164, Nantes Université-Polytech Nantes, 44306 Nantes, France
2. Department of Mathematics, Laboratoire J. A. Dieudonné, Côte d'Azur University, CEDEX 2, 06108 Nice, France; rene.lozi@univ-cotedazur.fr
3. Department of Computer Science, Laboratoire d'Informatique, de Robotique et de Micro Electronique de Montpellier (LIRMM), University of Montpellier, UMR CNRS 5506, CEDEX 05, 34392 Montpellier, France; william.puech@lirmm.fr
* Correspondence: safwan.elassad@univ-nantes.fr

Citation: El Assad, S.; Lozi, R.; Puech, W. Special Issue on Cryptography and Its Applications in Information Security. *Appl. Sci.* **2022**, *12*, 2588. https://doi.org/10.3390/app12052588

Received: 21 February 2022
Accepted: 25 February 2022
Published: 2 March 2022

Publisher's Note: MDPI stays neutral with regard to jurisdictional claims in published maps and institutional affiliations.

Copyright: © 2022 by the authors. Licensee MDPI, Basel, Switzerland. This article is an open access article distributed under the terms and conditions of the Creative Commons Attribution (CC BY) license (https://creativecommons.org/licenses/by/4.0/).

1. Introduction

Nowadays, mankind is living in a cyber world. Modern technologies involve fast communication links between potentially billions of devices through complex networks (satellite, mobile phone, Internet, Internet of Things (IoT), etc.). The main concern posed by these entangled complex networks is their protection against passive and active attacks that could compromise public security (sabotage, espionage, cyber-terrorism) and privacy.

To face it, most of the world web traffic (digital multimedia contents such as images, speech signal, videos, and emails) is protected against security threats, occurring among different societies and within several societal levels. Even governments (rogue or not) and some of their official agencies are suspected of promoting and actively participating in the hacking of other government officials, democratic processes, industrial secrets, and the citizens.

Thousands of private or official hackers target the sensitive information of citizens, industries, and governments. The threat is actual, and it is escalating year after year.

The aim of this Special Issue on "Cryptography and its Applications in Information Security" was to address the range of problems related to the security of information in networks and multimedia communications and to bring together researchers, practitioners, and industrials interested by such questions. Papers both from theoretical and practical aspects were welcome, including ongoing research projects, experimental results, and recent developments related to, but not limited to, the following topics: cryptography; chaos-based cryptography; block and stream ciphers; hash functions; steganography; watermarking; selective encryption; multimedia data hiding and security; secure FPGA implementation for cryptographic primitives; security methods for communications; Wireless Network Security (Internet, WSNs, UMTS, WiFi, WiMAX, WiMedia, and others); sensor and mobile ad hoc network security; security and privacy in mobile systems, secure cloud computing; security and privacy in social networks, vehicular networks, Web services; database security and privacy; intellectual property protection, lightweight cryptography for green computing; personal data protection for information systems; protocols for security; cryptanalysis, side channel attack; fault injection attack; and physical layer security for communications.

2. The Papers

In this Special Issue, we received a total of 24 submissions and, after the peer review, accepted and published 8 outstanding papers that span across several interesting topics on security, relationship between chaos pseudo-random numbers and stream ciphers, and blockchain technologies.

In the field of security, four papers are presented. The first one suggests employing a homomorphic encryption (HE) scheme that can directly perform arithmetic operations on ciphertexts without decryption to protect the model parameters. Using the HE scheme, the proposed privacy-preserving federated learning (PPFL) algorithm enables a centralized server to aggregate encrypted local model parameters without decryption. Furthermore, the proposed algorithm allows each node to use a different HE private key in the same FL-based system using a distributed cryptosystem [1].

A second paper in this field proposes a new anomaly detection algorithm for the Intrusion Detection System (IDS), where a machine learning algorithm is applied to detect deviations from legitimate traffic, which may indicate an intrusion. It involves a novel approach based on the transformation of the network flow statistics to gray images on which the Gray-Level Co-occurrence Matrix (GLCM) is applied together with an entropy measure recently proposed in the literature—2D Dispersion Entropy. This approach is assessed using the recently public IDS data set CIC-IDS2017. The results show that it is competitive in comparison to other approaches proposed in the literature on the same data set [2].

The main objective of the third paper is the classification of the Strongly Asymmetric Public Key Agreement (SAPKA) algorithms. SAPKA is a class of key exchanges between Alice and Bob that was introduced in 2011. The greatest difference from the standard PKA algorithms is that Bob constructs multiple public keys and Alice uses one of these to calculate her public key and her secret shared key. Therefore, the number of public keys and calculation rules for each key differ for each user. Although algorithms with high security and computational efficiency exist in this class, the relation between the parameters of SAPKA and its security and computational efficiency has not yet been fully clarified. By attempting algorithm attacks, the authors found that certain parameters are more strongly related to security. On this basis, they construct concrete algorithms and a new subclass of SAPKA, in which the responsibility of maintaining security is significantly more associated with the secret parameters of Bob than those of Alice [3].

The last paper in security designs a secure chaos-based stream cipher (SCbSC) and evaluates its hardware implementation performance in terms of computational complexity and its security. The fundamental element of this system is the proposed secure pseudo-chaotic number generator (SPCNG). The architecture of the proposed SPCNG includes three first-order recursive filters, each containing a discrete chaotic map and a mixing technique using an internal pseudo-random number (PRN). The three discrete chaotic maps, namely, the 3D Chebyshev map (3D Ch), the 1D logistic map (L), and the 1D skew-tent map (S), are weakly coupled by a predefined coupling matrix M. The mixing technique combined with the weak coupling technique of the three chaotic maps allows the system to be protected against side-channel attacks (SCAs) [4].

Linked to the topic of this paper, two other papers analyze the performances of stream ciphers. In [5], the bit independence criterion, which was proposed to evaluate the security of the S-boxes used in block ciphers, is assessed and improved. This paper proposes an algorithm that extends this criterion to evaluate the degree of independence between the bits of inputs and outputs of the stream ciphers. The effectiveness of the algorithm is experimentally confirmed in two scenarios: random outputs independent of the input, in which it does not detect dependence; and in the RC4 ciphers, where it detects significant dependencies related to some known weaknesses. The complexity of the algorithm is estimated based on the number of inputs l, and the dimensions, n and m, of the inputs and outputs, respectively.

Alternatively, in [6], a novel intermittent jumping CML system based on multiple chaotic maps is proposed. The intermittent jumping mechanism seeks to incorporate the multi-chaos, and to dynamically switch coupling states and coupling relations, varying with spatiotemporal indices. Extensive numerical simulations and comparative studies demonstrate that, compared with the existing CML-based systems, the proposed system has a larger parameter space, better chaotic behavior, and comparable computational complexity. These results highlight the potential of the proposal for deployment into an image cryptosystem.

The third topic highlighted in this Special Issue is the blockchain theory, either for digital cash or "digital authorization" for museums. Digital cash is a form of money that is stored digitally. Its main advantage when compared to traditional credit or debit cards is the possibility of carrying out anonymous transactions. Diverse digital cash paradigms have been proposed during recent decades, providing different approaches to avoid the double-spending fraud, or features such as divisibility or transferability. In [7], a new digital cash paradigm that includes the so-called no-valued e-coins, which are e-coins that can be generated free of charge by customers, is proposed. This new paradigm has also proven its validity in the scope of privacy-preserving pay-by-phone parking systems, and the authors believe it can become a very versatile building block in the design of privacy-preserving protocols in other areas of research.

The American Alliance of Museums (AAM) recently stated that nearly a third of the museums in the United States may be permanently closed since museum operations are facing "extreme financial difficulties", especially since the outbreak of COVID-19 at the beginning of this year (2020). The research published in [8] aimed at museums using the business model of "digital authorization". It proposes an authorization mechanism based on blockchain technology protecting the museums' digital rights in the business model and the application of cryptography. The signature and time stamp mechanism achieve non-repudiation and a timeless mechanism, which combines blockchain and smart contracts to achieve verifiability, non-forgery, decentralization, and traceability, as well as the non-repudiation of the issue of cash flow with signatures and digital certificates, for the digital rights of museums in business.

Author Contributions: All the editors have contributed equally. All authors have read and agreed to the published version of the manuscript.

Funding: This research received no external funding.

Acknowledgments: This issue would not be possible without the contributions of the authors who submitted their valuable papers. We would like to thank all reviewers and the editorial team of Applied Sciences for their great work.

Conflicts of Interest: The authors declare no conflict of interest.

References

1. Park, J.; Lim, H. Privacy-Preserving Federated Learning Using Homomorphic Encryption. *Appl. Sci.* **2022**, *12*, 734. [CrossRef]
2. Baldini, G.; Ramos, J.L. Intrusion Detection Based on Gray-Level Co-Occurrence Matrix and 2D Dispersion Entropy. *Appl. Sci.* **2021**, *11*, 5567. [CrossRef]
3. Satoshi Iriyama, S.; Jimbo, K.; Regoli, M. New Subclass Framework and Concrete Examples of Strongly Asymmetric Public Key Agreement. *Appl. Sci.* **2021**, *11*, 5540. [CrossRef]
4. Dridi, F.; El Assad, S.; Youssef, W.E.H.; Machhout, M.; Lozi, R. The Design and FPGA-Based Implementation of a Stream Cipher Based on a Secure Chaotic Generator. *Appl. Sci.* **2021**, *11*, 625. [CrossRef]
5. Madarro-Capó, E.J.; Legón-Pérez, C.M.; Rojas, O.; Sosa-Gómez, G.; Socorro-Llanes, R. Bit Independence Criterion Extended to Stream Ciphers. *Appl. Sci.* **2020**, *10*, 7668. [CrossRef]
6. Huang, R.; Han, F.; Liao, X.; Wang, Z.; Dong, A. A Novel Intermittent Jumping Coupled Map Lattice Based on Multiple Chaotic Maps. *Appl. Sci.* **2021**, *11*, 3797. [CrossRef]
7. Ricard Borges, R.; Sebé, F. A Digital Cash Paradigm with Valued and No-Valued e-Coins. *Appl. Sci.* **2021**, *11*, 9892. [CrossRef]
8. Wang, Y.-C.; Chen, C.-L.; Deng, Y.-Y. Authorization Mechanism Based on Blockchain Technology for Protecting Museum-Digital Property Rights. *Appl. Sci.* **2021**, *11*, 1085. [CrossRef]

Article

Privacy-Preserving Federated Learning Using Homomorphic Encryption

Jaehyoung Park [1] and Hyuk Lim [2,*]

[1] School of Electrical Engineering and Computer Science, Gwangju Institute of Science and Technology (GIST), Gwangju 61005, Korea; jaehyoungpark@gist.ac.kr
[2] AI Graduate School, Gwangju Institute of Science and Technology (GIST), Gwangju 61005, Korea
* Correspondence: hlim@gist.ac.kr

Abstract: Federated learning (FL) is a machine learning technique that enables distributed devices to train a learning model collaboratively without sharing their local data. FL-based systems can achieve much stronger privacy preservation since the distributed devices deliver only local model parameters trained with local data to a centralized server. However, there exists a possibility that a centralized server or attackers infer/extract sensitive private information using the structure and parameters of local learning models. We propose employing homomorphic encryption (HE) scheme that can directly perform arithmetic operations on ciphertexts without decryption to protect the model parameters. Using the HE scheme, the proposed privacy-preserving federated learning (PPFL) algorithm enables the centralized server to aggregate encrypted local model parameters without decryption. Furthermore, the proposed algorithm allows each node to use a different HE private key in the same FL-based system using a distributed cryptosystem. The performance analysis and evaluation of the proposed PPFL algorithm are conducted in various cloud computing-based FL service scenarios.

Keywords: privacy preserving; homomorphic encryption; federated learning

Citation: Park, J.; Lim, H. Privacy-Preserving Federated Learning Using Homomorphic Encryption. *Appl. Sci.* **2022**, *12*, 734. https://doi.org/10.3390/app12020734

Academic Editors: Safwan El Assad, René Lozi and William Puech

Received: 14 December 2021
Accepted: 7 January 2022
Published: 12 January 2022

Publisher's Note: MDPI stays neutral with regard to jurisdictional claims in published maps and institutional affiliations.

Copyright: © 2022 by the authors. Licensee MDPI, Basel, Switzerland. This article is an open access article distributed under the terms and conditions of the Creative Commons Attribution (CC BY) license (https://creativecommons.org/licenses/by/4.0/).

1. Introduction

Artificial intelligence (AI) is a technology that enables machines to realize human learning and reasoning abilities. This technology has been rapidly advancing and playing a significant role in our daily lives. In AI technology, data acquisition is crucial because AI technologies require model training using a certain amount of data for reliable AI-based services, and the performance of AI-based services is considerably affected by the training data quality. However, there are difficulties in data collection because the data may contain sensitive private information. In order to overcome these difficulties, federated learning (FL), in which training is performed without sharing sensitive local data, has been proposed in [1]. In FL, a centralized server sends a global model for AI learning to many distributed devices, which return local model parameters to the centralized server after training the model with local data. The centralized server updates the global model parameters using the locally trained model parameters from the distributed devices and sends the updated global model parameters to the distributed devices. This procedure is repeated until convergence is achieved. FL has the advantage of preventing the leakage of sensitive private information because it does not require local data sharing. However, recent research has shown that the local data of distributed devices can be leaked through the trained local model parameters, and attackers can exploit this loophole to infer sensitive information on the FL participant in [2,3].

Homomorphic encryption (HE) is a technology that enables arithmetic operations on ciphertexts without decryption. Aono et al. utilized an HE scheme to protect local gradients trained with local data in [2]. Using the HE scheme, the centralized server can update the global model parameters with the encrypted local gradients based on the homomorphic operation. Therefore, the distributed devices participating in FL, which we refer to as clients, do not have to concern about data leakage through local gradients because they deliver encrypted local gradients to the server. However, the clients must share the same private key in the FL-based system because homomorphic operations can only be performed between values encrypted with the same public key. In FL-based systems where many distributed devices, such as smartphones and Internet of Things (IoT) devices participate, the same private key for decryption can be distributed to many clients. Suppose the same private key is shared with many clients. Then, the probability of the private key being leaked or a malicious participant accessing other participants' data increases, which can weaken privacy protections in FL-based systems. As the result, stealing one client's private key can nullify the data privacy protection of all clients participating in FL systems. To overcome this vulnerability, this paper proposes a privacy-preserving federated learning (PPFL) algorithm that allows a cloud server to update global model parameters by aggregating local parameters encrypted by different HE keys in the same FL-based system using homomorphic operations based on a distributed cryptosystem.

This paper is organized as follows. In Section 2, we present related works on FL and the privacy issues of FL. Section 3 describes the preliminaries for understanding the FL algorithm and the cryptosystem for homomorphic operations. In Section 4, we describe the system and attack models for the proposed PPFL algorithm. Next, Section 5 explains the proposed PPFL algorithm using the distributed cryptosystem based on an additive homomorphic encryption (AHE) scheme. Afterwards, Section 6 presents a theoretical analysis of the proposed PPFL algorithm, and Section 7 presents experimental results to verify the performance of the proposed PPFL scheme. Finally, Section 8 concludes the paper.

2. Related Work

FL is one possible solution for preserving privacy in the machine learning field because the clients participating in the training process deliver only local model parameters trained with local data to a centralized server. McMahan et al. demonstrated the feasibility of FL by conducting empirical evaluation in various FL scenarios in [1]. Since then, many studies have been conducted to improve FL performance for learning accuracy, fairness, robustness, security, and privacy in various environments, such as IoT, edge, and cloud computing in [4–6]. In [7], a lightweight federated multi-task learning framework was proposed to provide fairness among participants and robustness against a poisoning attack that reduces learning accuracy. In [8], an FL framework using device-to-device communication was proposed to overcome the degradation in energy and learning efficiency due to frequent uplink transmissions between participants and physically distant central servers.

The studies on FL can be classified according to how they collect and process data for FL. In a case where the data have the same feature space and a different sample space, it is classified as horizontal FL, and in a case where the data have a different feature space and the same sample space, it is classified as vertical FL in [9]. In vertical FL, data alignment must be performed for vertical data utilization by sharing several different feature spaces. In this process, privacy is not protected because row data exchange may be required. For preserving data privacy, HE-based vertical FL algorithms were implemented by utilizing a trusted third party [4,9–11]. An approach to collaboratively train a high-quality tree boosting model was proposed to simplify FL-based systems by omitting third parties and showed that the performance of the proposed scheme was as accurate as the performance of centralized learning techniques in [12]. Horizontal FL is an algorithm in which multiple devices train a learning model using local data with the same feature space and share the trained model data to train a global model, and the scheme presented in [1] was a representative horizontal FL. The horizontal FL can be implemented without a data alignment process because it has the same feature space. Although many studies have been conducted for the development of FL, privacy threats still exist in FL. It was shown that sensitive private data could be leaked through the local gradients in [2,13], and participants' data can be inferred through a generative adversarial network using the global and the local model parameters in [3].

Several studies have been conducted to solve the privacy issues associated with the local model parameters in an FL-based system in [2,13–15]. Shokri and Shmatikov proposed a privacy-preserving deep learning (PPDL) algorithm where several distributed participants collaborate to train a deep learning model using local data; they established a trade-off relationship between practicality and security for the number of clients participating in the training process in [13]. Moreover, Aono et al. suggested a PPDL algorithm that encrypts local model parameters using HE schemes to protect the local and global model parameters in [2]. In the algorithm proposed in [2], strict key management is required and reliable channels for conveying ciphertexts must be established because all participants use the same private key for HE. In [16], HE-based federated learning was proposed, and its overhead was analyzed. However, all clients participating in training still use the same key in the system. In [14], based on Shamir's t-out-of-n Secret Sharing in [17], they presented an algorithm that allows the server to perform updates using local model parameters containing noise that can be canceled out through the cooperation of the participants in an FL-based system. The scheme proposed in [14] can prevent leakage of local model parameters due to the noise contained in the local parameter but can be vulnerable to insider attacks because participants must actively cooperate. Recently, Xu et al. offered a technique in which participants verify the integrity of the updated results in the system that updates the global model parameters using the Secret Sharing scheme in [15].

The algorithms using the HE scheme in [2] and Secret Sharing in [14] have shown that neural networks can be safely trained without personal information leakage in FL scenarios. In [2], all training process participants owned the same private key, although the distributed deep learning system using the HE scheme was designed to protect shared data. For this reason, all channels between participants and servers must be protected using transport layer security or secure socket layer. However, as the number of participants increases, the cost to establish secure channels becomes very high. In addition, since the probability of one participant's private key being leaked is the same as the probability of all participant's private keys being leaked in this system, the risk of private key leakage increases as the number of participants grows. In the proposed system, each participant can own different private keys on the same FL-based system. In [14], at least half of the participants must guarantee their honesty for privacy preservation. If the number of participants in FL is large, this assumption is reasonable, but there can be a variety of FL scenarios. The proposed system allows participants to preserve data privacy regardless of the number of honest participants and can be utilized as a solution to build a flexible PPFL-based system.

This paper presents a PPFL algorithm based on a distributed cryptosystem using the AHE scheme to protect the local and global model parameters. The participant uses the HE scheme to encrypt the local model parameters with its private key, and the cloud server updates the global model parameters with the local model parameter encrypted with different keys based on the distributed cryptosystem. The proposed PPFL-based system can achieve robust privacy protection because the proposed algorithm can allow each node to use a different private key for the HE scheme in the same FL-based system. Furthermore, a highly flexible FL-based system can be built using our algorithm because clients only need to encrypt and decrypt model parameters to protect them.

3. Preliminary
3.1. Federated Learning

In FL, multiple distributed servers or devices with local data train a machine learning model without exchanging local data. Distributed servers or devices share only local model parameters obtained by training a global learning model delivered from a centralized server with local data, allowing them to participate in the training process without concern about data leakage. The centralized server aggregates locally trained model parameters to update the global model and delivers the updated global parameters to distributed servers or devices to perform the training process again. This procedure is repeated until convergence is achieved.

According to the data distribution characteristics, FL can be categorized into horizontal federated learning, vertical federated learning, and federated transfer learning in [9]. Horizontal and vertical FL algorithms are applied when local datasets have the same feature space and different sample spaces and when local datasets have different feature spaces and the same sample space, respectively. Federated transfer learning is applied to a scenario where the local datasets have varying features and minimal overlapping samples. In this case, federated transfer learning utilizes the transfer learning techniques in [18] for FL-based systems. We consider horizontal FL in this paper. In other words, we assume that the local datasets have the same feature space and different sample spaces, and we consider an FL scenario in which many clients, including smartphones and IoT devices, participate in the training process.

3.2. Homomorphic Encryption

HE is a form of encryption that allows third parties to perform arithmetic operations directly with ciphertexts, and the HE scheme can be utilized to develop PPML in fields where data privacy is important. First, partial homomorphic encryption (PHE) capable of only addition or multiplication was developed. For example, the property of the AHE scheme, which can only perform addition operations, is represented as follows in [19]:

$$D_{sk_i}(E_{pk_i}(m_1) \cdot E_{pk_i}(m_2)) = m_1 + m_2, \tag{1}$$

where $D_{sk_i}(\cdot)$ is a decryption function using a private key sk_i, $E_{pk_i}(\cdot)$ is an encryption function using a public key pk_i, and m_i is a plaintext. The cloud server can perform homomorphic addition operations without decryption using (1). Subsequently, fully homomorphic encryption (FHE) capable of both addition and multiplication operations was established in [20] to overcome the limitations of PHE, which is challenging to implement various homomorphic operations in [21]. FHE enables a variety of operations to be implemented using addition and multiplication operations. These HE technologies have led to the development of PPML algorithms in the cloud and machine learning fields.

3.3. Distributed Homomorphic Cryptosystem

Distributed homomorphic cryptosystem (DHC) is a cryptosystem that can perform various homomorphic operations using secure multiparty computation (SMC) for implementing various homomorphic operations in a distributed manner. Figure 1 illustrates the decryption process of DHC. In typical public-key cryptography, parties with public and private keys perform encryption and decryption for secure communication, respectively. On the other hand, a private key is divided into several partial private keys in the DHC, and the partial private keys are distributed to multiple distributed servers. Distributed servers with partial private keys perform partial decryption using values encrypted with the public key. The other distributed server can obtain the plaintext by collecting the partially decrypted ciphertexts. This decryption process enables a variety of homomorphic operations based on multilateral cooperation.

The functions for the DHC are described as follows:

- **Key generation**: Function that generates a public-private key pair (pk_i, sk_i), $i \in \{1, \cdots, N_c\}$ of a user for given two large prime numbers p and q, where N_c is the number of clients participating in the local training. The public key is calculated by $p \cdot q$, and the private key corresponding to the public key is calculated by $lcm(p-1, q-1)/2$, where $lcm(x, y)$ denotes the least common multiple (LCM) of x and y. Note that the key size K is $p \cdot q$. Then, as the selected prime numbers increase, the computational complexity of cryptosystem increases because the complexity for the exponentiation operation of encryption and decryption increases [22]. Then, partially private keys

$[psk_i^{(1)}, psk_i^{(2)}, \cdots, psk_i^{(N_s)}]$ for distributed servers can be obtained by splitting the private key sk_i, where N_s is the number of distributed servers [23]. We select δ that satisfies $\delta \equiv 0 \bmod sk_i$ and $\delta \equiv 1 \bmod K^2$ at the same time and select y random numbers $\{a_1, a_2, \ldots, a_y\}$ from $\mathbb{Z}^*_{sk_i K^2}$. Then, we use these values to define the polynomial $p(x) = \delta + \sum_{i=1}^{y} a_i x^i$. The partial private key $psk_i^{(j)}$ is obtained by calculating the polynomial $p(x_j)$ using a non-zero value x_j from $\mathbb{Z}^*_{sk_i K^2}$.

- **Encryption**: Function that generates a ciphertext $E_{pk_i}(m) \in \mathbb{Z}^*_{K^2}$ for a plaintext $m \in \mathbb{Z}_K$, using a public key pk_i, where the key size K is $p \cdot q$. For simplicity, the ciphertext $E_{pk_i}(m)$ can be represented by $[m]_i$;
- **Decryption**: Function that decrypts a ciphertext $[m]_i$ using a private key sk_i and returns m;
- **Partial decryption**: Function that generates a partially decrypted ciphertext by partially decrypting $[m]_i$ using a partial private key $psk_i^{(k)}$, $k \in \{1, \cdots, N_s - 1\}$, as shown in Figure 1. For simplicity, the partially decrypted ciphertext $PD_{psk_i^{(k)}}([m]_i)$ can be denoted by $PD^k([m]_i)$;
- **Combined decryption**: Function that obtains and returns m using $(N_s - 1)$ partially decrypted ciphertexts $PD^k([m]_i)$ for $\forall k \in \{1, \cdots, N_s - 1\}$ and the partially private key $psk_i^{(N_s)}$. Note that a vector $\boldsymbol{PD}([m]_i)$ signifies $[PD^1([m]_i), PD^2([m]_i), \cdots, PD^{N_s-1}([m]_i)]$ in Figure 1.

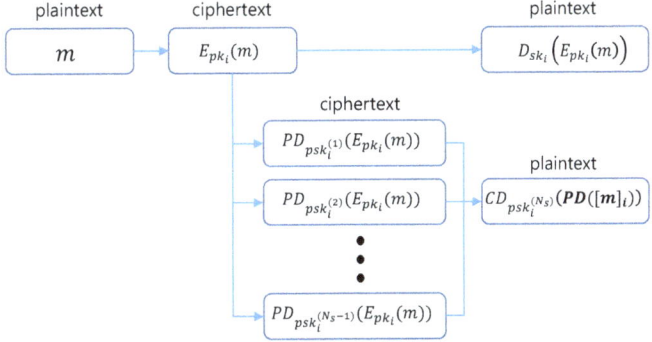

Figure 1. Diagram for encryption, decryption, and partial decryption.

Using the DHC, we have established a PPFL-based system in which the parties can jointly perform a global model update based on homomorphic operations to preserve the data privacy of the participants in the FL training process. The proposed PPFL algorithm is explained in detail in Section 5.

4. System and Attack Models
4.1. System Model

We consider a horizontal FL scenario that operates in a cloud system using SMC. A large amount of local model parameters is exchanged between the cloud server and many clients. Our proposed system encrypts the local model parameters using the DHC as described in Section 3.3 to protect data privacy. When a client first participates in FL, the certified key generation center generates a private and public key pair for the client. The private is sent to the client, and the public key is distributed to the client and servers in the system through a secure channel. In addition, after generating the private key, the certified key generation center splits the key into as many partial private keys as the number of the authenticated cloud server and computation provider, and distributes them to the servers one by one through a secure channel. If the private keys are stolen during the private key delivery process, the entire system may collapse. In the proposed system, secure channels for the private key delivery are built using secure sockets layer or transport layer security protocol. Since the private key delivery is performed intermittently, the possibility of an attacker stealing the private key is extremely low in the system. The cloud server and computation provider collaborate with each other to update the global model using the model parameters encrypted with different private keys from clients. Once the cloud server receives a set of model parameters from a client, it adds a random noise encrypted with the client's public key to the set of model parameters, partially decrypts it with the client's partial private key, and delivers it to the computation provider. The computation provider server obtains the partial decrypted sets of model parameters for the clients and decrypts them using the other partial private keys of the clients. Finally, the computation provider performs the model aggregation and encrypts it with the public key for each client, and returns it to the cloud server. The cloud server removes the random noise from the encrypted global model and sends it to each client. The detailed update process is described in Section 5.3.

Figure 2 depicts a simple system model comprising a key generation center (KGC), cloud server (CS), computation provider (CP), and multiple clients. The KGC is a trusted organization that performs authentication procedures for clients and servers and generates key pairs. The CS is responsible for securely combining the trained parameters on the clients and can select clients at every iteration for FL. The CP communicates directly with the CS and provides computational resources for requests of the CS. A single CP or multiple CPs can exist in the system, and the CPs and the CS perform cooperative encryption described in Section 3.3. Clients own each private key for decryption and perform local training with local datasets. In this system model, we make the following assumptions:

- The CS, CP, and the clients may attempt to abuse each others' data;
- Both the CS and CP are not simultaneously compromised by attackers;
- The CS and the CP do not cooperate to access client information.

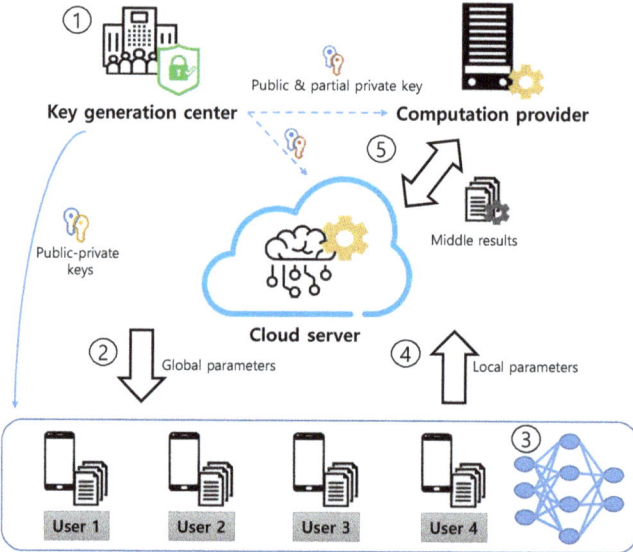

Figure 2. System model for privacy-preserving federated learning.

4.2. Attack Model

We consider several attack scenarios in the proposed system in terms of data privacy.

4.2.1. Single Malicious Entity Attack

- **Malicious clients**: Clients protect data privacy by encrypting shared parameters using the HE scheme. In the proposed system where all clients have different private keys, even if a malicious client can eavesdrop on all channels between the cloud and the clients, the malicious client cannot access the data because it cannot decrypt the ciphertexts without the corresponding private key. In addition, even if multiple malicious clients cooperate, they cannot access other clients' data without the corresponding private key;
- **Single malicious server**: The CS cannot access decrypted values in the proposed system because it only receives and handles encrypted values. For the CP, local model parameters can be accessed through combined decryption when partially decrypted values are delivered from the CS. However, the CS can prevent the CP from accessing the local model parameters by adding random noise to the encrypted local model parameters.

4.2.2. Cooperative Attacks by Multiple Malicious Entities

- **Malicious clients and CS**: If the CS cooperates with malicious clients, it can access the local model parameters of other clients because the CS can decide which clients participate in every iteration. For example, the CS can determine the list of participants with one client and the other malicious clients and calculate the average of encrypted local model parameters based on the HE scheme to obtain encrypted global model parameters. Then, the malicious clients can access the local parameters of the honest client by offsetting the local model parameter of the malicious clients by sharing their parameters because malicious clients can decrypt the global model parameter from the CS. This threat can be eliminated by ensuring more than one honest client at each iteration. In the proposed system, the threat can be eliminated by delivering the sum of local parameters through the cooperation of two or more honest clients. In addition, the privacy threat can be kept very low by ensuring the randomness of client selection through the KGC. When one honest client participates in the learning process, the conditional probability of all remaining participants being malicious clients can be expressed as $P_m^{N_c-1}$, where P_m is the ratio of the number of malicious clients to the total number of clients, and N_c is the number of clients participating in the local training. Thus, the probability of having access to local parameters of honest clients becomes very small, despite multiple malicious clients and CS collaborating. For example, even if half of all clients are malicious and N_c is 20, the probability is less than 2×10^{-6}.
- **Malicious clients and CP**: The CP cannot access the client information because random noise is added to the client information by the CS. Even if several malicious clients cooperate with the CP, the CP cannot access the local model parameters because the CS samples random noise for each client.
- **Malicious CS and CP**: When the CS and the CP cooperate to access a shared local model parameter, the client's information may be leaked. If all the distributed servers participating in the secure aggregation algorithm of SMC are compromised and cooperate with each other, there is no way to protect personal information. This paper assumes that the CS and the CP may be compromised simultaneously but do not cooperate to access client data. These assumptions are needed for building SMC-based DHC. To improve security in practice, we can increase the number of CPs participating in the secure aggregation algorithm, reducing the probability that multiple CPs and CS are malicious servers that cooperate with each other. In addition, an authentication procedure for the distributed cloud servers can be performed at the KGC to guarantee the servers participating in SMC are honest.

5. Privacy Preserving Federated Learning

In the proposed PPFL system, each client participating in the training process encrypts local model parameters trained with local data, using its own private key to protect the trained local model parameters. Thereafter, the clients transmit the encrypted local model parameters to the CS. The CS updates the global model parameters with the local model parameters encrypted with different keys by exploiting the partial homomorphic decryption capabilities of CPs. As a result, the proposed PPFL algorithm ensures data confidentiality between the CS and the clients, as well as data confidentiality among the clients because each client has its own private key and does not send the private key to other third parties. The detailed procedure of the proposed PPFL is described in the following subsections.

5.1. Homomorphic Key Generation and Distribution

As shown in step ① of Figure 2, individual public-private key pairs (pk_i, sk_i) for $i \in \{1, \cdots, N_c\}$ are generated at the KGC and are sent to clients for encryption and decryption through secure channels, where N_c is the number of clients participating in the training process of the proposed system. Before the KGC distributes the key pairs, it performs an authentication procedure for the clients and delivers the public-private key pairs to authenticated clients. The clients' public keys for encryption and a list of authenticated clients are transmitted to the CS and the CP, and the CS utilizes only local model parameters from authenticated clients. In addition, the partially private keys $[psk_i^{(1)}, psk_i^{(2)}, \cdots, psk_i^{(K)}]$ generated by the KGC are only sent to the CS and the CPs through secure channels for cooperative decryption, respectively.

5.2. FL Local Model Training

The CS selects a machine learning model to be trained on the client's side using local data. A deep neural network model is selected; however, other machine learning models can also be used for the proposed PPFL algorithm. The CS determines the percentage of clients participating in the training process and randomly selects clients to participate in the actual training process. At the first iteration, the CS encrypts the initial global model parameters using the selected client's public keys and sends the encrypted global model parameters to the clients, as shown in step ②. In the following iterations, the CS sends the results of aggregating the local model parameters using homomorphic operations to the clients without additional encryption because the result of the homomorphic operation is also an encrypted value. The global model parameter vector encrypted with the public key of the i-th client c_i is represented as $[W_g]_i$, where W_g is a global model vector containing the global model parameters.

The i-th client decrypts the encrypted global model vector, $[W_g]_i$, using its own private key and uses the decrypted global model parameters for the local training process. Each client participating in the proposed PPFL performs the training process using the local data in a deep neural network initialized with the global model parameters, as shown in step ③. After the local training process, the i-th client obtains local model parameters and proceeds to encrypt a local model vector W_l^i containing the local model parameters using its own public key. The client sends the encrypted local model vector to the CS for secure aggregation, as described in step ④. Note that the local model vector encrypted with the public key of the i-th client is represented as $[W_l^i]_i$. After the CS receives the encrypted local model vector from the clients, the encrypted vectors are used to update the global model vector through cooperation with the CP.

5.3. Secure Global Model Update with DHC

We propose a secure averaging local model vector algorithm that updates the global model vector by calculating the average of the local model vectors received from the clients. The CS utilizes the cooperative decryption scheme to obtain the average local model vector encrypted with different public keys in the same FL-based system through cooperation with the CP, as shown in step ⑤ of Figure 2.

Figure 3 illustrates the procedure of cooperative decryption and secure local model vector updates. First, the client sends an encrypted local value $E_{pk_i}(m)$ to the CS. The CS adds the encrypted random variable $E_{pk_i}(r)$ to the received ciphertext using homomorphic addition, where r is a random integer number, and then the CS can obtain $E_{pk_i}(m+r)$. The CS then forwards the ciphertext to the CPs. The (N_s-1) CPs perform partial decryptions, and the other CP performs the combined decryption to obtain the sum of the local value and random noise $(m+r)$. As explained in Figure 3, the CP can calculate the average of the sum of the local value and random noise $(m_{ave}+r_{ave})$ when receiving the sum from multiple clients. The sum's average is encrypted and sent back to the CS. Finally, the CS can remove the average of random values from the sum's average through homomorphic addition and obtain the encrypted average local value $E_{pk_i}(m_{ave})$ since the CS has the random values.

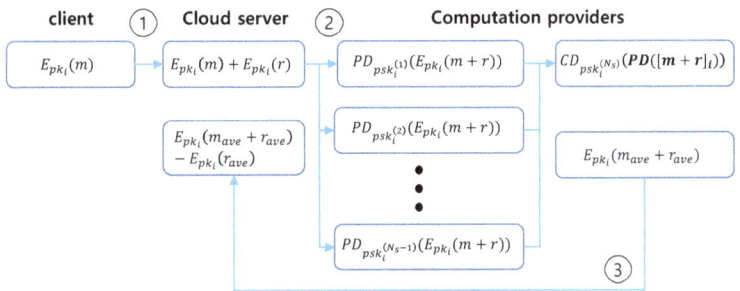

Figure 3. Diagram for secure averaging local model vector algorithm.

Algorithm 1 describes the proposed secure averaging local model algorithm where one CS and one CP exist. In the proposed algorithm, local model parameters encrypted with different keys are input, and global model parameters encrypted with different keys are output. The CS and CP have the partial private keys $psk_i^{(1)}$ and $psk_i^{(2)}$, respectively. The detailed procedure of the secure averaging local model vector algorithm is as follows:

1. The CS receives the encrypted local model vectors from the clients participating in the training process. Note that the local model vector encrypted with the public key of the i-th client is represented as $[W_l^i]_i$. Thereafter, the CS generates N_c random vectors with the same size as the local model vector and encrypts them using the client's public key, as shown in lines 2–3 of Algorithm 1;
2. The CS performs homomorphic addition operations with the encrypted local model vectors using the encrypted random vectors in line 5. The result of homomorphic addition between $[W_l^i]_i$ and $[R_i]_i$ is represented as $[S_i]_i$. Then, the CS partially decrypts the result of the homomorphic addition using the partial private key $psk_i^{(1)}$ in line 6. This process is repeated for N_c local model vectors. Subsequently, the CP sends the partially decrypted vectors $[PD^1([S_1]_1), PD^1([S_2]_2), \cdots, PD^1([S_{N_c}]_{N_c})]$ to the CP, as shown in lines 8–9;
3. As shown in lines 10–11, the CP partially decrypts the partially decrypted vectors using the partial private key $psk_i^{(2)}$ and obtains $[(W_l^1+R_1), \cdots, (W_l^{N_c}+R_{N_c})]$. Afterwards, the CP adds all the decrypted vectors and divides the sum by the number of clients N_c to obtain a vector containing the average parameters in line 12. The result is represented as W_{sum};
4. The CP encrypts W_{sum} using the public keys of the N_c clients in lines 13–15 and sends the encrypted vectors $[[W_{sum}]_1, [W_{sum}]_2, \cdots, [W_{sum}]_{N_c}]$ to the CS in line 16;
5. Finally, the CS calculates the encrypted average global model vectors for the clients by performing the homomorphic addition operation with the encrypted sum of random noises $[\frac{\sum_k -R_k}{N_c}]_i$, as shown in lines 17–19.

Algorithm 1 Secure averaging local model algorithm

1: Input: $[W_l^1]_1, [W_l^2]_2, \ldots, [W_l^{N_c}]_{N_c}$.
2: (@CS) generates N_c random vectors $R_1, R_2, \ldots, R_{N_c}$ and encrypts the random vectors
3: using the public keys.
4: **for** $i \leq N_c$ **do**
5:    (@CS) $[S_i]_i \leftarrow [W_l^i]_i \cdot [R_i]_i$.
6:    (@CS) Partially decrypts $[S_i]_i$ using $psk_i^{(1)}$
7: **end for**
8: (@CS) Sends partially decrypted vectors $[PD^1([S_1]_1), PD^1([S_2]_2), \ldots, PD^1([S_{N_c}]_{N_c})]$
9: to the CP.
10: (@CP) Partially decrypts the partially decrypted values using $psk_i^{(2)}$ and obtains
11: $[(W_l^1 + R_1), \ldots, (W_l^{N_c} + R_{N_c})]$.
12: (@CP) Calculates $W_{sum} = \frac{\sum_k W_l^k + \sum_k R_k}{N_c}$.
13: **for** $i \leq N_c$ **do**
14:    (@CP) Encrypts W_{sum} using the public key pk_i.
15: **end for**
16: (@CP) Sends the encrypted values $[[W_{sum}]_1, [W_{sum}]_2, \ldots, [W_{sum}]_{N_c}]$ to the CS.
17: **for** $i \leq N_c$ **do**
18:    (@CS) $[W_g]_i \leftarrow [W_{sum}]_i \cdot [\frac{\sum_k - R_k}{N_c}]_i$.
19: **end for**
20: Output: global weight vectors for the clients $[[W_g]_1, [W_g]_2, \ldots, [W_g]_{N_c}]$.

After updating the global model vector by performing the proposed secure averaging local model vector algorithm, the CP sends the updated global model vector to the clients for the next federated round. The clients execute the local training process using the updated global model vector as shown in Section 5.2 after decrypting the encrypted global model vector using its own private key. Thereafter, they send the newly trained local model vector to the CS, and then the CS and CP work together to update the global model vector. These procedures are repeated until convergence is achieved.

5.4. Data Structure and Protocol

The HE scheme increases data security but has the disadvantage of incurring communication overhead. Especially since the data length after encryption is independent of the plaintext length to be encrypted, the communication efficiency is significantly reduced if only one parameter is encrypted and sent. The proposed system establishes a data structure that can transfer multiple parameters to alleviate this efficiency degradation. The data structure for a weight w_i consists of a bit representing the sign, a zero bit to prevent an overflow caused by homomorphic additions, and the remaining bits signifying the weight's value. The number of weights included in one ciphertext can be calculated as $D = \lfloor \frac{K}{L_o} \rfloor$, where $\lfloor \cdot \rfloor$ is a round-down operation and L_o is the data length used for representing weights. Then, the data format to be encrypted can be expressed as follows: $[w_D^{Lo \cdot (D-1)}, w_{D-1}^{Lo \cdot (D-2)}, \ldots, w_1]$. Furthermore, since the secure aggregation operations are performed in the plaintext space except for the process of adding noise, the proposed algorithm can be implemented using only homomorphic operations for integer processing. Therefore, in the proposed system, integer numbers are used for data transmission, and they are represented as floating-point numbers using a decimal point pre-agreed between the clients and the servers after decryption.

6. Performance Analysis
6.1. Computational Overhead
6.1.1. Computational Overhead on the Client's Side

In the proposed PPFL algorithm, additional encryption operations are performed to protect the trained local model parameters, and extra decryption operations are performed to reflect the global model parameters to the learning model on the client's side. In the PCK scheme used in the proposed algorithm, the exponentiation operation has a dominant effect on encryption and decryption. The exponentiation operation g^r requires $1.5 \times N_r$ multiplications, where g is a generator of order $(p-1)(q-1)/2$, $r \in \mathbb{Z}_K$ is a random number, and N_r is the length of r in the DHC scheme in [21]. Thus, the computational complexity of the encryption operation in the proposed PPFL algorithm is given as $\mathcal{O}(N_r \cdot N_w)$, where N_w is the number of elements of the local model vector. Similarly, the computational complexity of the decryption in the proposed PPFL algorithm is also represented as $\mathcal{O}(N_r \cdot N_w)$.

6.1.2. Computational Overhead on the Server's Side

The computational complexity of the averaging local model parameter algorithm performed in the conventional FL algorithm can be expressed as $\mathcal{O}(N_w \cdot N_c)$. On the other hand, in the proposed PPFL algorithm, additional encryption, partial decryption, and homomorphic addition operations are required to perform the proposed secure averaging local model vector algorithm on the server's side. The encryption and partial decryption operations have a dominant impact on the computational complexity because the exponentiation operation in the encryption and partial decryption requires much more computation than the other operations. Moreover, as the number of clients and model parameters increases, the number of encryption and partial decryption operations to be performed also grows. Thus, the computational complexity of the secure averaging local model vector algorithm on the server's side can be represented as $\mathcal{O}(N_r \cdot N_w \cdot N_c)$.

6.2. Communication Overhead

6.2.1. Communication Overhead between Clients and the Cloud Server

In an FL scenario involving many clients, the communication overhead has a tremendous impact on performance. If a cryptosystem is utilized to preserve data privacy in an FL-based system, the communication overhead may be more significant than sending local parameters as a plaintext. In this paper's cryptosystem, the length of the ciphertext is affected only by the key size, regardless of the length of data the client sends to the server, and the length of data must be less than the key size. Thus, the closer the data length is to the key size, the less communication overhead is incurred because more information can be conveyed in one ciphertext. In the cryptosystem, since encryption requires a modular operation with a dividend K^2, the length of the ciphertext becomes $K \times 2$ bits. When the key size is K bits and the length of data to be transmitted is L_d bits, the transmission data volume after encryption becomes $2N/L_d$ times larger. As L_d is closer to K, the transmission data volume is approximately doubled. In the proposed system, K-bit data representing multiple local model parameters are generated to reduce the communication overhead.

6.2.2. Communication Overhead between the Cloud Server and the Computation Provider

In order to perform the proposed secure aggregation operation for local model vectors, we exploit the partial homomorphic decryption capabilities of CPs. The CS sends partially decrypted ciphertexts to the CP and receives ciphertexts from the CP in the proposed PPFL-based system. The length of the partially decrypted ciphertext is also $2 \times K$ bits in [21]. As shown in Algorithm 1, the amount of information communicated between the CS and the CP increases as the number of clients and model parameters increases. Thus, the communication overhead between the CS and the CP can be represented as $\mathcal{O}(N_c \cdot N_w)$ bits.

6.3. Overhead Comparison

Compared with the PPDL system in [2] that used the HE scheme, the proposed technique requires additional overhead to allow clients to use different private keys. In the proposed system, a certain degree of computational overhead for cooperative decryption and encryption processes is added, and the communication overhead is also added for the data exchanges between the CS and the CPs. The computational and communication overhead analysis was performed in Sections 6.1.2 and 6.2.2, respectively. The computational overhead comparison between the PPDL system and the proposed system is shown in Section 7.1, and the communication overhead between the CS and the CP is shown in Table 1.

6.4. Security Analysis

In the proposed algorithm, since model data is protected by the cryptosystem, attackers cannot access the data even if data are stolen from a communication network. Therefore, the attackers must break the cryptosystem to access local model data. Even in the case of insider attacks, attackers have to break the cryptosystem to access the data because clients have access only to global models and their local models, and servers have access only to encrypted data. The HE scheme can have a higher security level of cryptosystem if the key size increases. As the key size increases, the amount of computation required to break the encryption algorithm or system also increases. For example, it was shown that if the key sizes are 1024, 2048, 3072, 7680, and 15,360 bits in Paillier's cryptosystem-based HE scheme, the security level is given as 80, 112, 128, 192, and 256-bit, respectively in [24]. However, there exists a trade-off relationship between the security level and computation/communication overhead because the amount of encryption and decryption computation and the data length of the ciphertexts also increase. Numerical evaluations of computation and communication overheads with respect to the key size have been performed in Section 7.

With a higher security level of the cryptosystem, the proposed scheme can more robustly resist the attacks described in Section 4.2.

- In the attack model of malicious clients, malicious clients can eavesdrop on the communication channels, and obtain ciphertexts and partially decrypted ciphertexts. Because the DHC is semantically secure, as described in [25], an attacker has to break the cryptosystem to obtain private data. If an honest client uses a longer key size, the attacker will have to use more computational resources to break the victim's cryptosystem.
- In the attack model of a single malicious server, malicious servers can eavesdrop on the communication channels, and obtain ciphertexts and partially decrypted ciphertexts. As in the client attack model, an attacker must break the cryptosystem to obtain private data. Even though CP acquires the model parameters and performs the combined decryption to obtain the plaintext, it cannot access private data because the plaintext includes a random noise added by the CS.
- In the attack model of malicious clients and servers, the malicious entities can cooperate; the malicious client may provide a private key to the malicious server. If all clients use the same private key, the malicious server can access all clients' private data because a malicious client can provide the private key to the malicious server. On the other hand, since the proposed system allows clients to use different private keys in the same FL-based system, the privacy leakage can be prevented in this attack model.

Based on the observation of the attack models, it is worth noting that the proposed system provides a much stronger level of security than the state-of-the-art system proposed in [2] where all clients use the same private key. Even though a client's cryptosystem is broken, the data privacy of the other clients in the proposed system is not affected by the compromised cryptosystem of the victim client because the clients use different private keys. Suppose that the amount of computational resources required to break the cryptosystem of the i-th client is $C_b^i(K)$, where K is the key size. If all clients use the same private key, the computational resource amount to break the system is expressed as $\min\{C_b^1(K), C_b^2(K), \ldots, C_b^{N_c}(K)\}$. As a result, in this case, if the most vulnerable client is broken in, the entire system can be easily compromised. On the other hand, the computational resource amount to break the proposed system is given by $\sum_i^{N_c} C_b^i(K)$. As the number of clients increases, the amount of computational resources needed to attack the system increases linearly.

7. Performance Evaluation

In this section, we have developed the proposed algorithm using Python and evaluated the performance on a workstation (3.6 GHz quad-core processor and 8 GB RAM) in terms of computation and communication overhead.

7.1. Computational Overhead

Figure 4 shows the running time measured for performing encryption and decryption according to the key size. In our simulation environment, the key sizes were selected as 1024, 2048, 3072, 7680, and 15,360 bits to achieve 80, 112, 128, 192, and 256-bit security levels, respectively. For example, the encryption took 11.7, 78.4, and 1552.5 ms for 80, 128, and 265 bit security, respectively, in Figure 4. As the key size increases, the running times for encryption and decryption increase exponentially because the exponent of the exponentiation operation in encryption and decryption increases.

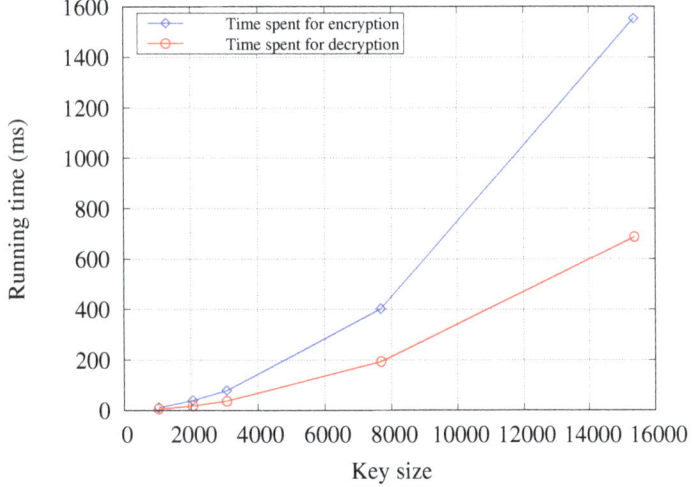

Figure 4. Running time to execute homomorphic encryption and decryption with respect to the key size.

Figure 5 shows the running times measured for performing the proposed secure averaging local model algorithm in Algorithm 1 with respect to the number of clients in the cryptosystem with different key sizes. The convolutional neural network with 105,506 parameters was used for the simulation study, and the data length used for representing weights was set to 16 bits. As the number of clients increases, the number of homomorphic operations increases as the number of parameters to protect using the HE scheme increases, and thus the running time increases linearly. In Figure 5, the running time increases as the key size increases. In fact, if the key size is larger, the total number of ciphertexts to be delivered is smaller. However, as shown in Figure 4, the running time of homomorphic operations increases exponentially as the key size increases. As a result, the total running time of Algorithm 1 increases as the key size increases. Nevertheless, as the key size increases, the security level of the system increases. This is because the higher the key size, the greater the number of cases is required to break the cryptosystem. Thus, because the computational burden and security gain have a trade-off relationship, we can select an appropriate key size according to system requirements.

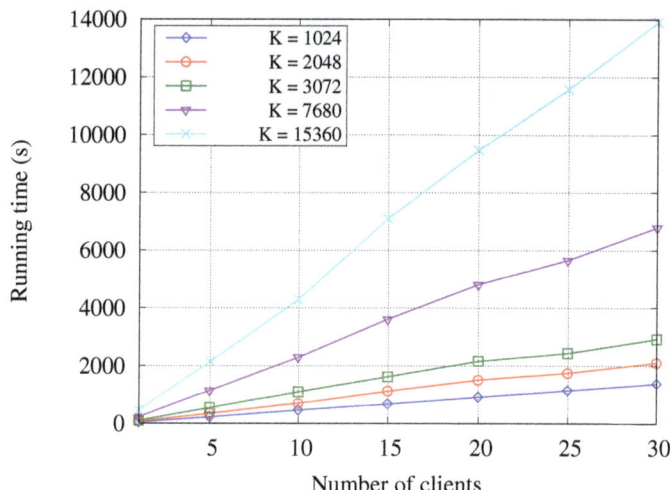

Figure 5. Running time to execute the proposed secure averaging local model algorithm with respect to the number of clients in the cryptosystem with different key sizes.

Figure 6 shows the running time for performing the proposed algorithm with respect to the neural network size of the federated learning. The number of clients was set to 10, and the data length used for representing weights was set to 16 bits for the simulation study. As the number of parameters increases, the running time increases because the amount of information to be processed by the homomorphic operation increases.

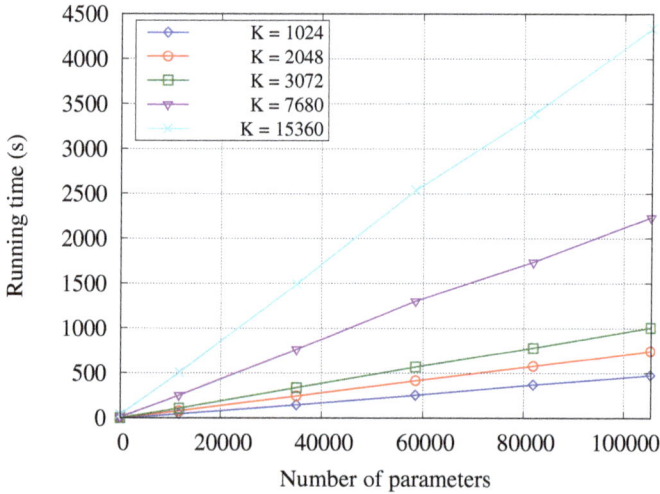

Figure 6. Running time to execute the proposed secure averaging local model algorithm with respect to the neural network size in the cryptosystem with different key sizes.

We performed simulations to compare the computational overhead of the proposed PPFL system and the PPDL system proposed in [2]. The key size K is 1024, and the number of parameters is 105,506 in the simulation environment. In Figure 7, it is seen that the computation overhead of the proposed system is about 2.3 times greater than that of the PPDL system. This is because the additional encryption and partial decryption processes are performed at the servers to make clients have different keys in the proposed system. Despite the greater computational overhead of the proposed algorithm, the security intensity of FL systems is significantly improved because the clients use different private keys. Therefore, the proposed system can be deployed in a more adversarial environment where there exist many malicious clients and they are difficult to be identified. In future work, we will research how to reduce the overhead in PPFL while retaining the same strong security level.

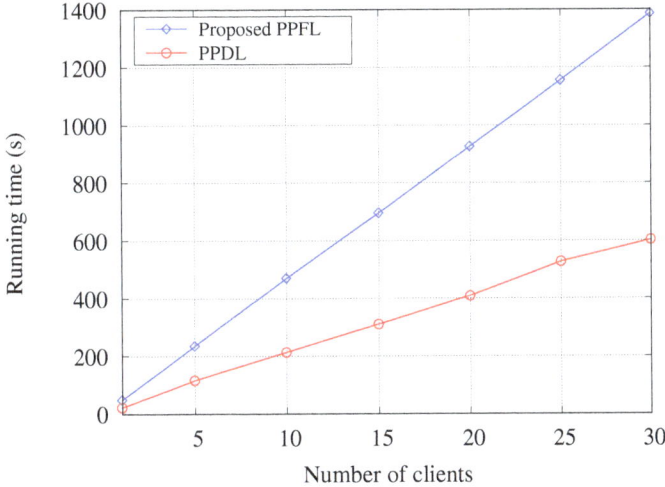

Figure 7. Running time to execute the proposed PPFL and the PPDL using the Paillier cryptosystem with respect to the number of clients.

7.2. Communication Overhead

The communication overhead increases in the proposed PPFL algorithm because the servers and clients communicate with each other using the encrypted model vectors to protect the model parameters. Table 1 shows the communication overhead with respect to the key size when the number of parameters is 105,506, and the data length for one parameter is 16 bits. The communication overhead remains almost constant regardless of the key size as shown in Table 1. If the key size increases, the ciphertext length may become longer, but since the number of parameters included in the ciphertext increases, the key size has little effect on the communication overhead of the proposed algorithm. In addition, as the number of clients increases, the communication overhead linearly increases because the amount of data exchanged between CS and CP increases.

Table 1. Communication overhead (KB).

K (Key Size)	1024	3072	15,360
Client–CS	422.1	422.4	422.4
CS–CP ($N_c = 10$)	8443	8448	8448
CS–CP ($N_c = 100$)	84,429	84,480	84,480

8. Conclusions and Future Work

This paper has proposed the PPFL system based on the HE scheme to protect shared model parameters in an FL-based system. Furthermore, we have proposed a technique for the secure aggregation of local model parameters encrypted with different keys in the same FL-based system. In the proposed system, the computational and communication costs required to improve security level in FL were theoretically analyzed, and the performance of the proposed PPFL algorithm in terms of overhead was evaluated via simulations. In the future, our research focuses on how to further reduce the computation and communication costs in the proposed PPFL algorithm while retaining privacy preservation of clients, and also focuses on how to determine an appropriate number of clients participating in FL to expedite the learning and to reduce latency of FL-based services.

Author Contributions: Conceptualization, J.P. and H.L.; methodology, J.P.; investigation, J.P.; formal analysis, J.P. and H.L.; validation, J.P. and H.L.; writing—original draft preparation, J.P.; writing—review and editing, H.L.; and supervision, H.L. All authors have read and agreed to the published version of the manuscript.

Funding: This work was supported by Institute of Information and Communications Technology Planning and Evaluation (IITP) grant funded by the Korea government (MSIT) (No. 2021-0-00379, Privacy risk analysis and response technology development for AI systems).

Institutional Review Board Statement: Not applicable.

Informed Consent Statement: Not applicable.

Conflicts of Interest: The authors declare no conflicts of interest.

References

1. McMahan, B.; Moore, E.; Ramage, D.; Hampson, S.; y Arcas, B.A. Communication-efficient learning of deep networks from decentralized data. In Proceedings of the International Conference on Artificial Intelligence and Statistics, Fort Lauderdale, FL, USA, 20–22 April 2017; pp. 1273–1282.
2. Aono, Y.; Hayashi, T.; Wang, L.; Moriai, S. Privacy-preserving deep learning via additively homomorphic encryption. *IEEE Trans. Inf. Forensics Secur.* **2017**, *13*, 1333–1345.
3. Hitaj, B.; Ateniese, G.; Perez-Cruz, F. Deep models under the GAN: Information leakage from collaborative deep learning. In Proceedings of the ACM SIGSAC Conference on Computer and Communications Security, Dallas, TX, USA, 30 October–3 November 2017; pp. 603–618.
4. Zhang, C.; Xie, Y.; Bai, H.; Yu, B.; Li, W.; Gao, Y. A survey on federated learning. *Knowl.-Based Syst.* **2021**, *216*, 106775. [CrossRef]
5. Khan, L.U.; Saad, W.; Han, Z.; Hossain, E.; Hong, C.S. Federated learning for internet of things: Recent advances, taxonomy, and open challenges. *IEEE Commun. Surv. Tutor.* **2021**, *23*, 1759–1799. [CrossRef]
6. Lin, J.C.W.; Srivastava, G.; Zhang, Y.; Djenouri, Y.; Aloqaily, M. Privacy-preserving multiobjective sanitization model in 6G IoT environments. *IEEE Internet Things J.* **2020**, *8*, 5340–5349. [CrossRef]
7. Li, T.; Hu, S.; Beirami, A.; Smith, V. Ditto: Fair and robust federated learning through personalization. In Proceedings of the International Conference on Machine Learning, Virtual, 18–24 July 2021; pp. 6357–6368.
8. Lin, F.P.C.; Hosseinalipour, S.; Azam, S.S.; Brinton, C.G.; Michelusi, N. Semi-decentralized federated learning with cooperative D2D local model aggregations. *IEEE J. Sel. Areas Commun.* **2021**, in press. [CrossRef]
9. Yang, Q.; Liu, Y.; Chen, T.; Tong, Y. Federated machine learning: Concept and applications. *ACM Trans. Intell. Syst. Technol. (TIST)* **2019**, *10*, 1–19. [CrossRef]
10. Ou, W.; Zeng, J.; Guo, Z.; Yan, W.; Liu, D.; Fuentes, S. A homomorphic-encryption-based vertical federated learning scheme for rick management. *Comput. Sci. Inf. Syst.* **2020**, *17*, 819–834. [CrossRef]
11. Zhang, C.; Li, S.; Xia, J.; Wang, W.; Yan, F.; Liu, Y. Batchcrypt: Efficient homomorphic encryption for cross-silo federated learning. In Proceedings of the 2020 USENIX Annual Technical Conference (USENIXATC 20), Online, 15–17 July 2020; pp. 493–506.
12. Cheng, K.; Fan, T.; Jin, Y.; Liu, Y.; Chen, T.; Papadopoulos, D.; Yang, Q. Secureboost: A lossless federated learning framework. *IEEE Intell. Syst.* **2021**, in press. [CrossRef]
13. Shokri, R.; Shmatikov, V. Privacy-preserving deep learning. In Proceedings of the ACM SIGSAC Conference on Computer and Communications Security, Denver, CO, USA, 12–16 October 2015; pp. 1310–1321.
14. Bonawitz, K.; Ivanov, V.; Kreuter, B.; Marcedone, A.; McMahan, H.B.; Patel, S.; Ramage, D.; Segal, A.; Seth, K. Practical secure aggregation for privacy-preserving machine learning. In Proceedings of the ACM SIGSAC Conference on Computer and Communications Security, Dallas, TX, USA, 30 October–3 November 2017; pp. 1175–1191.
15. Xu, G.; Li, H.; Liu, S.; Yang, K.; Lin, X. Verifynet: Secure and verifiable federated learning. *IEEE Trans. Inf. Forensics Secur.* **2019**, *15*, 911–926. [CrossRef]
16. Fang, H.; Qian, Q. Privacy Preserving Machine Learning with Homomorphic Encryption and Federated Learning. *Future Internet* **2021**, *13*, 94. [CrossRef]
17. Shamir, A. How to share a secret. *Commun. ACM* **1979**, *22*, 612–613. [CrossRef]
18. Pan, S.J.; Yang, Q. A survey on transfer learning. *IEEE Trans. Knowl. Data Eng.* **2009**, *22*, 1345–1359. [CrossRef]

19. Paillier, P. Public-key cryptosystems based on composite degree residuosity classes. In Proceedings of the International Conference on the Theory and Application of Cryptology and Information Security, Singapore, 14–18 November 1999; pp. 223–238.
20. Gentry, C. Fully homomorphic encryption using ideal lattices. In Proceedings of the Annual ACM Symposium on Theory of Computing, Bethesda, MD, USA, 31 May–2 June 2009; pp. 169–178.
21. Liu, X.; Deng, R.H.; Choo, K.K.R.; Weng, J. An efficient privacy-preserving outsourced calculation toolkit with multiple keys. *IEEE Trans. Inf. Forensics Secur.* **2016**, *11*, 2401–2414. [CrossRef]
22. Katz, J.; Lindell, Y. *Introduction to Modern Cryptography*; CRC Press: Boca Raton, FL, USA, 2020.
23. Liu, X.; Choo, K.K.R.; Deng, R.H.; Lu, R.; Weng, J. Efficient and privacy-preserving outsourced calculation of rational numbers. *IEEE Trans. Dependable Secur. Comput.* **2018**, *15*, 27–39. [CrossRef]
24. Barker, E.; Barker, E.; Burr, W.; Polk, W.; Smid, M. *Recommendation for Key Management: Part 1: General*; National Institute of Standards and Technology, Technology Administration: Gaithersburg, MD, USA, 2006.
25. Bresson, E.; Catalano, D.; Pointcheval, D. A simple public-key cryptosystem with a double trapdoor decryption mechanism and its applications. In Proceedings of the International Conference on the Theory and Application of Cryptology and Information Security, Taipei, Taiwan, 30 November–4 December 2003; Springer: Berlin/Heidelberg, Germany, 2003; pp. 37–54.

Article

Intrusion Detection Based on Gray-Level Co-Occurrence Matrix and 2D Dispersion Entropy

Gianmarco Baldini [1,*], Jose Luis Hernandez Ramos [1] and Irene Amerini [2]

[1] European Commission, Joint Research Centre, 21027 Ispra, Italy; jose-luis.hernandez-ramos@ec.europa.eu
[2] Department of Computer, Control and Management Engineering A. Ruberti, Sapienza University of Rome, 00185 Rome, Italy; amerini@diag.uniroma1.it
* Correspondence: gianmarco.baldini@ec.europa.eu; Tel.: +39-334-2300960

Citation: Baldini, G.; Hernandez Ramos, J.L.; Amerini, I. Intrusion Detection Based on Gray-Level Co-Occurrence Matrix and 2D Dispersion Entropy. *Appl. Sci.* **2021**, *11*, 5567. https://doi.org/10.3390/app11125567

Academic Editors: Safwan El Assad, Arcangelo Castiglione, René Lozi and William Puech

Received: 20 April 2021
Accepted: 11 June 2021
Published: 16 June 2021

Publisher's Note: MDPI stays neutral with regard to jurisdictional claims in published maps and institutional affiliations.

Copyright: © 2021 by the authors. Licensee MDPI, Basel, Switzerland. This article is an open access article distributed under the terms and conditions of the Creative Commons Attribution (CC BY) license (https://creativecommons.org/licenses/by/4.0/).

Abstract: The Intrusion Detection System (IDS) is an important tool to mitigate cybersecurity threats in an Information and Communication Technology (ICT) infrastructure. The function of the IDS is to detect an intrusion to an ICT system or network so that adequate countermeasures can be adopted. Desirable features of IDS are computing efficiency and high intrusion detection accuracy. This paper proposes a new anomaly detection algorithm for IDS, where a machine learning algorithm is applied to detect deviations from legitimate traffic, which may indicate an intrusion. To improve computing efficiency, a sliding window approach is applied where the analysis is applied on large sequences of network flows statistics. This paper proposes a novel approach based on the transformation of the network flows statistics to gray images on which Gray level Co-occurrence Matrix (GLCM) are applied together with an entropy measure recently proposed in literature: the 2D Dispersion Entropy. This approach is applied to the recently public IDS data set CIC-IDS2017. The results show that the proposed approach is competitive in comparison to other approaches proposed in literature on the same data set. The approach is applied to two attacks of the CIC-IDS2017 data set: DDoS and Port Scan achieving respectively an Error Rate of 0.0016 and 0.0048.

Keywords: intrusion detection systems; security; machine learning; communication

1. Introduction

Our society is becoming increasingly dependent on the internet and communication services but the risk of cybersecurity threats has also increased. Intrusion Detection System (IDS) can be a powerful tool to mitigate cybersecurity attacks. Research in IDS is more than 20 years old and various types of IDS have been proposed in literature: signature-based IDS, which focuses on the recognition of traffic patterns associated to a threat, anomaly-based IDS which detects deviations from a model of legitimate traffic and often relies on machine learning or reputation-based IDS based on the calculation of reputation scores [1]. Requirements or preferred features of IDS have been already defined in literature [1,2] and they can be summarized in: (a) fast detection of the attack, (b) high detection accuracy and (c) low computing complexity of the detection algorithm to support the capability to analyze a large amount of traffic due to the high throughput of the current networks. The successful fulfillment of these three main requirements can be challenging because there are trade-offs between them. For example, algorithms, which are able to obtain high detection accuracy, may require considerable computing resources or they may not be able to achieve a fast detection. The advantage of anomaly-based IDS, in comparison to signature-based IDS, is to potential detect new attacks which have not been recorded before and where the corresponding signature has not been created yet. On the other side, the detection of anomalies in high throughput traffic would benefit from dimensionality reduction while preserving an high detection accuracy. To achieve this goal, anomaly-based IDS have been proposed in literature where a sliding window is used [2,3].

This paper focuses on an anomaly detection approach where the network flows data is collected in windows of fixed size, which are then converted to gray images on which the Gray level Co-occurrence Matrix (GLCM) is calculated. Then, the features (e.g., contrast) of the GLCM are used as an input to a machine learning algorithm for the threat detection. In addition, the 2D Dispersion Entropy (2DDE) recently introduced in [4] is also calculated as additional feature of the GLCM. To the knowledge of the authors, this approach is novel in IDS literature both from the point of view of the application of GLCM and the application of 2D Dispersion Entropy. The application of the sliding window and the GLCM allows a significant dimensionality reduction. First of all, the number of samples of the data set is reduced by the size of the sliding window (W_S in the rest of this paper). For example, the data from the IDS is processed in windows of size $W_S = 100 *$ number of features of the data set ($N_F = 78$ for the data set used in this paper). Then, the window data is converted to a grayscale image, which implies a further dimensionality reduction because the output of GLCM is a matrix of size $Q_F * Q_F$ where Q_F is the quantization factor of GLCM. Then, the GLCM features (e.g., contrast, Shannon entropy) plus the 2DDE applied to GLCM is calculated to implement an additional dimensionality reduction step. Finally, the reduced data set is provided as an input to a machine learning algorithm. The application of the Sequential Feature Selection (SFS) algorithm (a wrapper feature selection algorithm) further reduces the number of features. The challenge is to preserve the discriminating characteristics in the data set, which allows to detect with significant accuracy the attack.

The rationales for the approach proposed in this paper are following: the first reason is related to the choice of using the GLCM beyond the need for dimensionality reduction as explained above. The idea is that the sequential structure of the network flows, in case of an intrusion, is altered in comparison to the legitimate traffic. Since the GLCM is created by calculating how often pairs of pixels with a specific value and offset occur in the image, the underlying idea of the approach is that numbers of pairs of pixels will be altered when an attack is implemented. Such changes will be reflected in the frequencies of the number of pairs, which (in turn) will have an impact on GLCM features (e.g., contrast) or information theory measures like entropy. The second reason for the proposed approach is that the classical Shannon entropy measure is only based on the histogram of GLCM elements while it would also be valuable to evaluate the sequences of GLCM elements since they may provide further information on the presence of the attack. For this reason, the 2D Dispersion Entropy (2DDE) was introduced in the study. As described in Section 3.4 later in this paper, 2DDE allows to analyze irregularity of images on the basis of the frequency of patterns in the image, which can provide more information than the classical Shannon entropy.

This study uses the CIC-IDS2017 data set [5], which has been recently published (2017) and it has been increasingly used by the IDS research community.

The results shown in this paper demonstrate that this approach manages to remain competitive in terms of detection performance in comparison to more sophisticated and computing demanding approaches based on Deep Learning (DL) applied to the same data set [6,7].

To summarize, the contributions of this paper are following:

- GLCM is applied to an IDS problem where the network traffic features are transformed to grayscale images on which GLCM is applied. An extensive evaluation of the GLCM hyperparameters on detection accuracy is implemented. To the knowledge of the authors this is the first time that the GLCM in combination with 2DDE is used for the IDS problem. This is also the first time that the authors submitted this study for review and the authors did not publish this work before.
- 2D Dispersion Entropy (2DDE) is used as additional GLCM feature. We demonstrate that the use of this entropy measure contributes significantly to the capability of the proposed approach to detect a cybersecurity attack.
- The study uses the recent IDS CIC-IDS2017 data set instead of older data sets, which may not be representative any longer of modern networks.

We highlight that the approach is based only on the network flow features and it does not attempt to perform a deep-packet inspection on the network traffic. In addition, it is limited in scope to two specific attacks of the CIC-IDS2017 data set: DDoS and Port Scan attack since they are the ones with the most significant number of samples in the data set and they are the ones where the research community has given much attention [7–10], which is relevant for the comparison of the results of this paper with literature (see Section 4).

The structure of this paper is the following: Section 2 provides the literature review. Section 3 describes the overall workflow of the approach, the concept of GLCM, the definition of 2D Dispersion Entropy and the materials (i.e., CICIDS2017 data set) used to evaluate the approach. In addition, Section 3 describes the machine learning algorithms adopted for the detection and the evaluation metrics. Section 4 presents the results, including the findings from the hyperparameters optimization phase and the comparison to the other approaches used in literature. Finally Section 5 concludes this paper.

2. Related Works

IDS have been proposed in literature for more than 20 years. As described in [1], IDS performs the essential function to detect unauthorized intruders and attacks to a wide scope of electronic devices and systems: from computers, to network infrastructures, ad-hoc networks an so on. From that seminal survey, many different types of IDS have been proposed and various classifications of IDS can be found in literature. One early classification in [1] defines two main IDS categories: offline IDS where the analysis of logs and audit records is performed some time after the traffic network operation (e.g., the analysis is executed the day after the network or computer system activity) and the online (or real-time) IDS where the analysis is performed directly on the traffic or immediately after the traffic features are calculated (e.g., average duration of the packets or average time of the connection). For example, the online IDS performs the analysis on a single or a set of observations (e.g., network flows) at the time after an initial training phase, while the offline IDS analyzes all the observations of the day before. More recent surveys like [11–13] provide different taxonomies for IDS. For example, IDSs can be classified in the category of signature detection or anomaly detection. In signature detection, the intrusion is detected when the system or network behavior matches an attack signature stored in the IDS internal databases. Signature-based IDSs have the advantage that they can be very accurate and effective at detecting known threats, and their mechanism is easy to understand. On the other side, signature-based IDSs are ineffective to detect new attacks and variants of known attacks, because a matching signature for these attacks is still unknown. In anomaly detection, the activities of a system at an instant (e.g., an observation or a set of observations of network traffic) are compared against the normal behavior profile calculated in a training phase against legitimate traffic. Machine Learning (ML) or DL can be used to evaluate how traffic samples are different from legitimate traffic and they can be used to classify the network traffic in the proper category. The disadvantages of the anomaly detection approach are the significant computing effort, the difficulty to define the proper model and the potential high number of False Positives (FP) [14].

The method proposed in this paper is anomaly detection, where a dimensionality reduction is performed to improve the detection time and accuracy. The dimensionality reduction is implemented using a sliding window approach where the initial data samples (the network flows data) are collected in windows of size W_S (this is the name of the parameter used in the rest of this paper). Then, features are calculated on the window set of data. This approach has been already used in literature to achieve dimensionality reduction [2,3]. In the rest of this section, we identify some key studies with a specific focus on IDS approaches based on the sliding window concept and/or the use of entropy measures. We also report on studies where image-based approaches are used in combination with ML or DL.

Shannon entropy is usually adopted as a feature calculated on the windowed set of data. The reason is that intrusion attacks have been demonstrated to alter the entropy of the network flows traffic. For example, the authors in [15] have proposed a detection method called D-FACE to differentiate legitimate traffic and DDoS attacks. The method compares the Shannon entropy calculated on the source IP data of the normal traffic flows with the traffic in a specific time window (e.g., the observation). This entropy difference is called Information Distance (ID) and is used as the detection metric when the calculated entropy goes beyond thresholds based on legitimate traffic. In another example, the authors of [10] have used a sophisticated approach to evaluate the difference between legitimate traffic and anomalous traffic potentially linked to a DDoS attack by using Shannon entropy. Then, the authors employ a Kernel Online Anomaly Detection (KOAD) algorithm using the entropy features to detect input vectors that were suspected to be DDoS. Another IDS approach based on sliding window and conditional entropy is proposed in [16] where anomalies related to various attacks including DDoS are detected in a two steps approach. The maximum entropy method is first used to create a normal model in which the classes of network packets are distributed and have the best uniform distribution. In a second step, conditional entropy is then applied to determine the difference between the distribution of packet classes in current traffic compared to the distribution found as a result of the maximum entropy method. The authors in [17] have also used a sliding window approach combined with Shannon entropy to detect Denial of Service Router Advertisement Flooding Attacks (DSRAFA). A fixed sliding window of 50 packets was used and a threshold mechanism was adopted to identify traffic anomalies which could indicate the attack.

The data presented in a sliding window can also be transformed to enhance the detection accuracy. With the advent of DL and Convolutional Neural Network (CNN) in particular, an approach adopted by some authors is to convert the batch data of a sliding window into an image, which is then provided as an input to a CNN based detection algorithm. This approach is proposed recently in [18] where the data of the network traffic flows is transformed to images which are given as input to CNN combined with Long Short Term Memory (LSTM). A similar approach is adopted in this paper with the difference that DL is not used since it can be quite time-consuming and a more conventional texture analysis approach is used together with a novel entropy measure. Another DL approach is proposed by the authors in [19] where a conditional variational autoencoder is used for intrusion detection in IoT. The conversion of flow features to grayscale images is also adopted in [20] where the authors propose a method which extracts 256 features from the flow and maps them into 16 ∗ 16 grayscale images, which are then used in an improved CNN to classify flows. On the other side, none of the papers investigated by the authors adopt other tools for image analysis for IDS like the GLCM adopted in this study. This may be due to the consideration that DL has become state of art in image processing even if it comes at the cost of a significant computational effort.

Then, the approach presented in this paper combines the image-based concept of [18,20] where the set of network flows are combined in images following the studies [10,15] where an information theory approach (e.g., entropy measure) is used in combination with conventional machine learning. We show in the Results Section 4 that this approach manages to provide competitive detection results in a time efficient way in comparison to studies using the same CICIDS2017 data set used in this paper.

3. Methodology and Materials

3.1. Workflow

The overall workflow of the proposed approach is shown in Figure 1 where the main phases are identified with numbers. The phases are described in the following bullet list:

1. The network flows for the labeled legitimate traffic are collected in a sliding window (the windows are not overlapping) of size W_S using all the 78 network features present in CICIDS2017 data set (see Section 3.2). Different sizes W_S of the sliding window are used in this study: $W_S = (100, 200, 300, 400, 500)$ network flows.

2. The sliding window data (of size $W_S * 78$) is converted to gray images by rescaling the values of the network flows features. The rescaling is implemented by converting the original values of the network flows in the sliding window to the range 0–256 (for each network flow feature) to obtain 256 levels of gray. A linear conversion is used. Examples of the resulting gray images for the Legitimate traffic and the Port Scan traffic are shown in Figure 2, where the y-axis represents the id of the network flow feature, while the x-axis represents the flow id. The sliding window is applied in sequential order regardless of the IP origin as it was created in the public data set [5] used in this paper.
3. The GLCM is applied to the gray images with different values of the GLCM hyparameters. See Section 3.3 for the definition of GLCM and hyperparameters. One of the important hyperparameters is the quantization factor Q_F. In other words, different GLCMs are created for each of the distances and directions considered in Section 3.3 (even for different values of G_D) and for the value of the quantization factor Q_F. The resulting size of the GLCM is $Q_F * Q_F$.
4. The GLCM features (e.g., contrast) are calculated. In addition, the 2DDE is also calculated on the images. The definition of the 2DDE is presented in Section 3.4.
5. The ML algorithm is applied to the features calculated in the previous step. The description of the algorithm used in this study and the related hyperparameters are described in the Section 3.5.

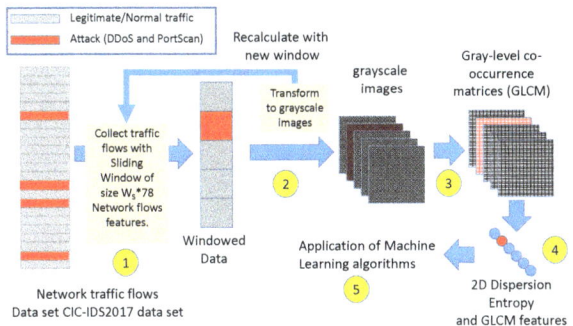

Figure 1. Overall workflow.

Finally, the hyperparameters of the GLCM and of the ML algorithm are tuned using the Error Rate (ER) as evaluation metric. The definition of ER and the other evaluation metrics are provided in Section 3.6.

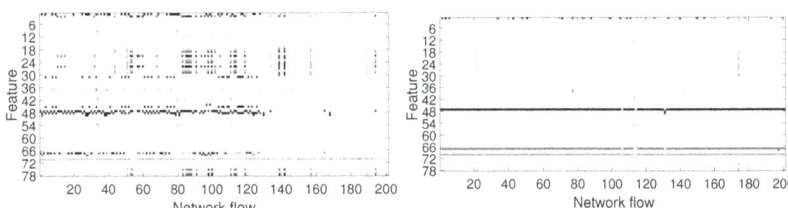

(**a**) Grayscale image of the Legitimate network flows for $W_S = 200$. (**b**) Grayscale image of the Port Scan network flows for $W_S = 200$.

Figure 2. An example of the grayscale images with $W_S = 200$ for legitimate and Port Scan network flows.

3.2. Materials

To evaluate the proposed approach, the publicly available CICIDS2017 data set described in [5] is used. This data set was used because it is relatively recent in comparison to older data set like the KDD-99 data set, whose limitations are known and discussed in [14,21]. These limitations have prompted the research community to generate even simulated data sets like the ones proposed in [22]. The CICIDS2017 data set is based on a real network where intrusion attacks have been implemented. Then, it satisfies one of the requirements for data sets identified in [21]. As described in [5], the test bed to implement the attacks was divided into two completely separated networks: a Victim-Network and the Attack-Network. In the Victim-Network, the creators of the CICIDS2017 data set have included routers, firewalls, switches, along with the different versions of the common three operating systems: Windows, Linux and Macintosh. The Attack-Network is implemented by one router, one switch and four PCs, which have the Kali and Windows 8.1 operating systems. The Victim-Network consists three servers, one firewall, two switches and ten PCs interconnected by a domain controller (DC) and active directory. The dataset contains normal traffic (i.e., legitimate traffic with no attacks) and traffic with the most up-to-date common attacks for five days. We selected two types of attacks in this study: the DDoS attack and the PortScan attack. These attacks are chosen because they are quite representative of intrusion attacks and because they have the largest number of samples in the CICIDS2017 data set. Both attacks were generated on the last day of the data set. The DDoS traffic in this dataset was generated with a tool to flood UDP and TCP requests to simulate network layer DDoS attacks, and HTTP requests to simulate application-layer DDoS attacks. The Portscan attack was executed from all the Windows machines by the main switches. The dataset is completely labeled and includes 78 network traffic features, which were extracted using the CICFlowMeter software package described in [5]. Note that the CICflowmeter outputs 84 features including the label (see [23] for a description of all the features), but we removed features 1 (Flow Id), 2 (Source IP), 3 (Source Port), 4 (Destination IP), 5 (Destination Port) thus obtaining the 78 features used in this paper, since the last field is used as the label.

Two separate data sets are created from the original CICIDS2017 data set: one data set containing only the legitimate traffic and the Distributed Denial of Service (DDoS) network flows and another data set containing only the legitimate traffic and the Port Scan network flows. The two data sets were created by selecting from the whole data set only the network flows labelled as legitimate traffic and the specific attack: DDoS or PortScan. All the network flows from the other attacks were removed from the other data set.

Table 1 shows the number of legitimate/benign traffic samples and the attack samples for the DDoS and the PortScan attacks.

Table 1. Number of samples for the legitimate/benign traffic and the DDoS and PortScan attacks in the CICIDS2017 data set considered in the study.

Attack	Number of Legitimate Samples	Number of Attack Samples
PortScan	2,273,097	158,930
DDoS	2,273,097	128,027

As described in Section 3.5 later in this paper, the data set is subdivided in folds, which contain exclusive portions of the data set containing both legitimate traffic and traffic related to the intrusion attack. In this study, a number of folds equal to 3 was selected to ensure to have enough samples of the attack since the CICIDS2017 data set is unbalanced like many other intrusion data sets: the number of traffic samples related to the intrusion are usually much less than the legitimate traffic ones. The application of the approach proposed in this paper is applied separately to each fold and then the values are averaged. The optimization step is also performed on averaging the results from all the folds. This technique of subdividing the data set is one of the guidelines for the application of machine learning to intrusion detection problem as suggested in [21]. It is important to point out that, in our study, we use all the 78 network flow features of the data set and we do not perform a feature selection on the network flow features as other papers have attempted [5,24]. The reason is that feature selection is performed on the GLCM features instead and we wanted to conduct the analysis on the widest set of information from the initial data set. We want to limit the degrees of freedom in the problem by not performing a feature selection on the network flows features. This is a similar approach to other papers where all the 78 Network flows features are used [25]. Future developments may investigate the selection of specific network flow features even if this task can be quite time consuming with this approach.

3.3. Gray Level Co-Occurrence Matrices

The GLCM is a statistical method for examining texture that considers the spatial relationship of pixels. The GLCM functions characterize the texture of an image by calculating how often pairs of pixel with specific values and in a specified spatial relationship occur in an image, creating a GLCM. In this context, the network flows features are used to create a grayscale image (256 levels of gray) X of size W_S (where W_S is the size of the sliding window) for N_F (where N_F is equal to the number of features or 78). Then, the GLCM is created on this grayscale image by calculating how often a pixel with the intensity (gray-level) value i occurs in a specific spatial relationship to a pixel with the value j. Each element (i,j) in the resultant GLCM is simply the sum of the number of times that the pixel with value i occurred in the specified spatial relationship to a pixel with value j in the input image. The GLCM is characterized by a number of hyperparameters in its definition: the most important is the quantization factor Q_F or the number of levels. From the original 256 gray levels of the source image, the GLCM introduces a new number of gray levels, specified as an integer: the Q_F. This parameter is quite important because the number of gray-levels determines the size of the resulting GLCM. This means that regardless of the size ($W_S * N_F$) of the input grayscale image, the resulting GLCM image has size $Q_F * Q_F$. The trade-off is that a larger Q_F may increase the granularity of the features on which the ML is applied, thus potentially increasing the detection accuracy. On the other side of the coin, a larger GLCM (and greater values of Q_F) increases the time to calculate the GLCM features and 2D Dispersion Entropy (2DDE). Then the value of Q_F must be optimized. Another hyperparameter is the distance between a pixel of interest and its neighbor. It is possible to define not only the absolute distance among pixels but also the angle as shown in Figure 3. In the rest of this paper, the absolute distance is named G_D and each distance and angle is defined by a 2-tuple (e.g., [0 G_D]).

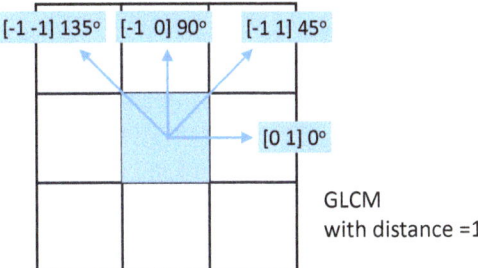

Figure 3. GLCM distance and angle parameter with $G_D = 1$.

A third parameter (Symmetric or Not Symmetric) is the order of values which can be counted only once or twice. When the hyperparameter is set to Symmetric the GLCM is calculated by counting the pairings twice. When the hyperparameter is set to Not Symmetric the GLCM is calculated by counting the pairings only once.

An example of the calculation of the GLCM applied to a grayscale image of size $4*5$ is provided in Figure 4 where $Q_F = 8$ and the distance/angle 2-tuple is set to [0 1].

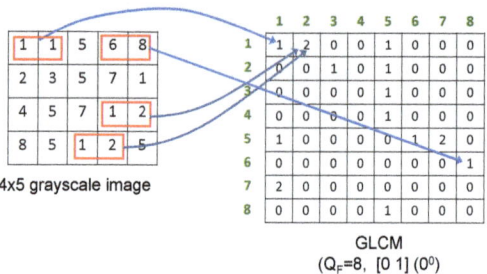

Figure 4. Example of the calculation of GLCM on the basis of a grayscale image.

In the original GLCM definition, it is possible to calculate the GLCM along all the possible directions, but an evaluation of the data set by the authors in this specific IDS context has shown that the additional directions not described in Figure 3 are duplications of the directions already identified and they would add unneeded computing efforts as they would grow the number of features on which the ML has to be applied. A quantitative confirmation that the angles shown in Figure 3 have an higher detection performance than using all the angles of the GLCM is provided in Section 4. Then, in the rest of this paper, we will use the 2-tuples [0 G_D], [$-G_D$ 0], [$-G_D$ $-G_D$], [$-G_D$ G_D].

As described in the Introduction Section 1, the idea to use GLCM in the context of IDS is that the sequential structure of the network flows in case of an intrusion is altered in comparison to the legitimate traffic. Since the GLCM is created by calculating how often pairs of pixels with a specific value and offset occur in the image, the underlying idea of the approach is that numbers of pairs of pixels will be altered when an attack is implemented. The challenge is that it is not known a priori how the choice of values of the hyperparameters influences the detection accuracy of the intrusion attack, since this information depends on the context (e.g., the topology of the network, the type of traffic and the type of attack). Then, an optimization process has to be performed, which is described in detail in Section 4.1.

3.4. Two dimensional Dispersion Entropy

Two dimensional dispersion entropy (2DDE) was introduced by Azami and others in [4] where it is described in detail. Here, we provide a brief description of the 2DDE measure with reference to the IDS problem.

2DDE is an extension of the one dimension dispersion entropy, which has demonstrated its superior performance in many problems [26]. The original definition of 2DDE is applied to an image of size $w*h$, but in this study, the GLCM is the image on which the 2DDE is to be calculated and its size is equal to $Q_F * Q_F$, then the equations and definitions from [4] are modified accordingly.

In a first step, each value in the image U (i.e., the GLCM image in this case) is mapped to classes with integer indices from 1 to c (which is one of the hyperparameters in the definition of 2DDE). To this aim, there are a number of linear and nonlinear mapping approaches, which can be used in the dispersion entropy based methods. The simplest and fastest algorithm is the linear mapping. However, when maximum or minimum values are noticeably larger or smaller than the mean/median value of the image (as in this case where anomalies significantly greater than the average must be detected), it is preferable to use a sigmoid function as defined in [4], where the normal cumulative distribution function (NCDF) is used to map the image into the classes, as this function naturally raises in a sigmoidal shape. The NCDF maps the initial image U (i.e., the GLCM of the window traffic) to Y with values from 0 to 1 as in the following equation:

$$y_{i,j} = \frac{1}{\sigma\sqrt{2\pi}} \int_{-\infty}^{x_{i,j}} e^{\frac{-(t-\mu)^2}{2\sigma^2}} dt \tag{1}$$

where μ and σ are the average and standard deviation of U.

The concept of Dispersion Entropy (even the one in one dimension) is related to the patterns of the embedding dimension m (another hyperparameter in the definition of 2DDE). The dispersion patterns are created in the following way.

First, a new matrix $z\, z_{k,l}^{m,c}$ is created from $y_{i,j}$ using the following equations (this is the adaption of Equation (2) from [4] taking in consideration that w and h from (2) are equal to Q_F in this case):

$$z_{k,l}^c = round(c \times y_{i,j} + 0.5) \tag{2}$$

where $z_{k,l}^c$ shows the (i,j)th of the classified image and rounding involves either increasing or decreasing a number to the next digit.

Second, $z_{k,l}^{m,c}$ are made with the embedding dimension vector according to the following equation.

$$\begin{aligned} z_{k,l}^{m,c} = &z_{k,l}^c, z_{k,l+1}^c, z_{k,l+2}^c, \ldots, z_{k,l+(m_{Q_F}-1)}^c, \\ &z_{k+1,l}^c, z_{k+1,l+1}^c, z_{k+1,l+2}^c, \ldots, z_{k+1,l+(m_{Q_F}-1)}^c, \ldots, \\ &z_{k+(m_{Q_F}-1),l}^c, z_{k+(m_{Q_F}-1),l+1}^c, z_{k+(m_{Q_F}-1),l+2}^c, \ldots, z_{k+(m_{Q_F}-1),l+(m_{Q_F}-1)}^c \end{aligned} \tag{3}$$

where $k, l = 1, 2, \ldots, Q_F - (m_{Q_F} - 1)$.

Third, each term of the matrix $z_{k,l}^{m,c}$ is mapped to a dispersion pattern π_{v_j} on which the final entropy measure 2DDE is calculated in the following way.

Fourth, for each $c^{m_{Q_F} \times m_{Q_F}}$ potential dispersion pattern $\pi_{v_0,v_1,\ldots,v_{(m_{QF-1})}}$, the relative frequency $\pi_{v_0,v_1,\ldots,v_{(m_{QF-1})}}$ in the image Y is calculated.

Finally, the Shannon entropy is calculated on the dispersion pattern $\pi_{v_0,v_1,\ldots,v_{(m_{QF-1})}}$ to provide the 2DDE according to the following equation:

$$2DDE(m,c) = -\sum_{\pi=1}^{c^{m_{Q_F} \times m_{Q_F}}} p\left(\pi_{v_0,v_1,\ldots,v_{(m_{QF-1})} \times (m_{Q_F}-1)}\right) \times ln\left(p\left(\pi_{v_0,v_1,\ldots,v_{(m_{QF-1})} \times (m_{Q_F}-1)}\right)\right) \tag{4}$$

As in one dimension dispersion entropy, the value of the parameters m and c should be tuned to achieve an optimal performance (in this case, the detection of the attack). On the other side, the parameters m and c are bound [4] by the size of the time series on which 2DDE has to operate. As described in [27] even for the one dimension dispersion entropy, to work with reliable statistics when calculating dispersion entropy, it is suggested that the number of potential dispersion patterns c^m is smaller than the length of the signal. In the two dimensional case, the rule reported in [4] and adapted for this case where the GLCM is square of size $Q_F * Q_F$, is that $(c^{m_{Q_F}})^2 < (Q_F - m_{Q_F} - 1)^2$, which limits the space of the values of m and c to few values as the range of Q_F. Considering that the range of Q_F spans from 12 to 48 in this study, the combinations of $c = 2, m = 3$ and $c = 3, m = 2$ are chosen. It must also be taken in consideration that higher values of m increase the computing time, which is not desirable for a large data set like the one used in this study.

As pointed out in the Introduction, the rational for using 2DDE in this study is that 2DDE allows to analyze irregularity of images on the basis of the frequency of the dispersion patterns in the image [4], which can provide more information than the classical Shannon Entropy. Since an intrusion attack usually disrupts the regularity of the structure of legitimate traffic, the application of 2DDE can provide a significant discriminating power for the detection of the attack.

3.5. Machine Learning Algorithms

The following machine learning algorithms were used in the study: the Support Vector Machine, the Decision Tree and Naive Bayes algorithm. These algorithms were chosen because they have already been used in literature [5,24] on the same problem because of their accuracy and cost effectiveness. These three algorithms are also chosen because each of them belongs to a specific category of machine learning algorithms and they are useful to provide a comparison on the relevance of the algorithm to the IDS problem. We would like anyway to remark that the goal of the paper is the investigation on the discriminating power of the approach based on GLCM and 2DDE rather than the choice of a specific machine learning algorithm. In other words, they are used rather to understand the relevance of the different features for the detection of benign and malicious activity, which can eventually serve as the basis for a non-machine-learning detector [14].

Support Vector Machine (SVM), is a supervised learning model which classifies data by creating a hyperplane or set of hyperplanes in a high dimensional space, to distinguish the samples belonging to different classes (two classes in this problem). Various kernels have been tried and the one providing the best performance was the Radial Basis Function (RBF) kernel, where the values of the scaling factor γ must be optimized together with the parameter C [28].

The Decision Tree algorithm is a predictive modeling approach where a decision tree (as a predictive model) analyzes the observations about an item (represented in the branches) to reach conclusions about the item's target value (represented in the leaves). In this case we use classification trees where leaves represent class labels and branches represent conjunctions of features that lead to those class labels. The hyperparameter chosen for optimization is the maximum number of branches at each split named N_B in the rest of this paper. It was chosen the option that the algorithm trains the classification tree learners without pruning them. The optimal values for the three machine learning algorithms are presented in Section 4.

The Naive Bayes (NB) machine learning algorithm is a probabilistic classifier, which is based on applying Bayes' theorem with strong (naïve) independence assumptions between the features [29]. In the NB algorithm, models assign class labels to problem instances, represented as vectors of feature values, where the class labels are drawn from some finite set. In many practical applications like IDS, the parameter estimation for the NB models uses the method of maximum likelihood; which means that the NB classifier can be applied even without accepting Bayesian probability.

As discussed before, for the application of all machine learning algorithms, a 3-fold approach (i.e., K-fold approach with K = 3) was used for classification, where 1/3 of the dataset was used for test, and 2/3 was used for training and validation. The portions of the data set in each fold are exclusive among themselves. The value of 3 was used to subdivide the data set in portions large enough to ensure that a meaningful set of data related to the intrusion is present in the input data to the classifiers. Then, the attack data was also split in 3 as part of the overall 3-fold approach. Since intrusion data sets are usually heavily unbalanced (legitimate traffic is much larger than traffic related to the intrusion), there is the risk that high values of K produce folds with a limited number of samples related to the attack. To further generalize the application of the approach, the overall classification process was then repeated 10 times, each time with different training and test sets. The final results were averaged.

As it is seen in the Section 4, the Decision Tree algorithm provides the optimal detection accuracy for this problem.

3.6. Detection Metrics

This subsection describes the metrics used to evaluate the performance of the approach proposed in this paper and the alternative approaches used in literature.

The main metric is the Error Rate (ER), which is 1-Accuracy and it is defined as:

$$\text{ER} = 1 - \frac{\text{TP} + \text{TN}}{(\text{TP} + \text{FP} + \text{FN} + \text{TN})} \quad (5)$$

where TP is the number of True Positives, TN is the number of True Negatives, FP is the number of False Positives and FN is the number of False Negatives.

To complement the accuracy metric, the True Positive Rate (TPR) and the False Positive Rate (FPR) are used, which are defined in the following equations:

$$\text{TPR} = \frac{\text{TP}}{(\text{TP} + \text{FN})} \quad (6)$$

$$\text{FPR} = \frac{\text{FP}}{(\text{FP} + \text{TN})} \quad (7)$$

Another method to evaluate the performance of the approach proposed in this paper, is the Receiver Operative Characteristics (ROC) curve which is created by plotting the True Positive Rate (TPR) against the False Positive Rate (FPR) at various threshold settings. A metric based on the ROC curve is the Equal Error Rate (EER), which is the point on the ROC curve that corresponds to have an equal probability of miss-classifying a positive or negative sample.

3.7. Features and Hyperparameters

As specified before, the proposed approach is based on a number of hyperparameters, which are summarized in the following bullet list with the related trade-offs:

- W_S = the size of the sliding window. The trade-off is that a small value of W_S does not provide an image large enough for the application of GLCM while a large value of W_S limits the number of samples for the application of ML.
- Q_F = Quantization factor in GLCM. A value too small may not provide enough granularity for an effective detection of the threat while a value, which is too large increases significantly the computing time.
- GLCM distance and angle parameter in GLCM definition. There are no trade-offs but the optimal value must be selected.
- GLCM symmetry. If the GLCM is applied with symmetry or asymmetry. There are no trade-offs but the optimal value must be selected.
- c and m in the 2DDE definition. c and m are bound by the value of Q_F as specified in [4].
- hyperparameters in the ML algorithm. For example, the maximum number of branches at each split in the Decision Tree algorithm.

Beyond the hyperparameters identified above, a number of features were proposed by Haralick in its seminal paper on the design of GLCM and the related feature [30,31], but not all the Haralick features are applicable to this context, either because they are computing intensive, because they are unstable for small images or because they are not relevant for the context. In addition, the use of all the Haralick Features in combination for the hyperparameters identified above would generate a search space which would be too large for the optimization process. After a preliminary assessment of the Haralick features, the following set of features were used for this study and they are listed in Table 2. The approach presented in this paper was to combine a pre-selected set of Haralick features in addition to 2DDE with the *GLCM symmetry* and *GLCM distance and angle parameter* hyperparameters. The other Haralick features described in [31] were not used because their detection performance using ER was suboptimal in comparison to the features identified in Table 2: Difference Variance, Difference Entropy, Info Measure of Correlation, Maximum Correlation Coefficient. Note that in Table 2, Energy indicates the angular second moment and Homogeneity is the inverse difference moment on the basis of the terms described in [31].

Table 2. List of features used in this study (G_D is the GLCM distance and the 2-tuples indicate the angles used to build the GLCM).

Feature Id F_{ID}	Feature Name and Parameters
1,17,33,49	Contrast [0 G_D], [−G_D 0], [−G_D −G_D], [−G_D G_D] Not Symmetric
9,25,41,57	Contrast [0 G_D], [−G_D 0], [−G_D −G_D], [−G_D G_D] Symmetric
2,18,34,50	Energy [0 G_D], [−G_D 0], [−G_D −G_D], [−G_D G_D] Not Symmetric
10,26,42,58	Energy [0 G_D], [−G_D 0], [−G_D −G_D], [−G_D G_D] Symmetric
3,19,35,51	Homogeneity [0 G_D], [−G_D 0], [−G_D −G_D], [−G_D G_D] Not Symmetric
11,27,43,59	Homogeneity [0 G_D], [−G_D 0], [−G_D −G_D], [−G_D G_D] Symmetric
4,20,36,52	Correlation [0 G_D], [−G_D 0], [−G_D −G_D], [−G_D G_D] Not Symmetric
12,28,44,60	Correlation [0 G_D], [−G_D 0], [−G_D −G_D], [−G_D G_D] Symmetric
5,21,37,53	Shannon Entropy [0 G_D], [−G_D 0], [−G_D −G_D], [−G_D G_D] Not Symmetric
13,29,45,61	Shannon Entropy [0 G_D], [−G_D 0], [−G_D −G_D], [−G_D G_D] Symmetric
6,22,38,54	2DDE ($m = 2, c = 3$) [0 G_D], [−G_D 0], [−G_D −G_D], [−G_D G_D] Not Symmetric
14,30,46,62	2DDE ($m = 2, c = 3$) [0 G_D], [−G_D 0], [−G_D −G_D], [−G_D G_D] Symmetric
7,23,39,55	2DDE ($m = 3, c = 2$) [0 G_D], [−G_D 0], [−G_D −G_D], [−G_D G_D] Not Symmetric
15,31,47,63	2DDE ($m = 3, c = 2$) [0 G_D], [−G_D 0], [−G_D −G_D], [−G_D G_D] Symmetric
8,24,40,56	Sum of variances [0 G_D], [−G_D 0], [−G_D −G_D], [−G_D G_D] Not Symmetric
16,32,48,64	Sum of variances [0 G_D], [−G_D 0], [−G_D −G_D], [−G_D G_D] Symmetric

As it is shown in Table 2 the GLCM is calculated on the gray image (created from the sliding window) for different values of the angle for a specific value of the distance G_D. Then, the related features for each specific GLCM are calculated. For example, one GLCM is calculated for the angle [0 G_D] while another GLCM is calculated for the angle [−G_D −G_D].

The optimal set of features are selected using the forward sequential feature selection. In the forward sequential search algorithm, optimal features are added to a candidate subset while evaluating the criterion. Since an exhaustive comparison of the criterion value at all subsets of the 64 features from Table 2 (repeated for all the values of the hyperparameters) is typically infeasible, the forward sequential search moves only in the direction of growing from an initial feature (the one with the lowest ER when all the features are considered). The best ten features are used to calculate the final metrics of evaluation: ER, FPR, FNR. The number of ten has been adopted because it was the optimal value between the need to limit the number of features for the application of ML and the increase of detection accuracy (beyond ten features, the improvement in detection accuracy was minimal).

Apart from the application of the ML algorithms, the difference of the discriminating power of 2DDE in comparison to the other features to detect an attack can be visualized by the trends of the features in comparison to the network flows features. Figure 5a,b show respectively the trend of GLCM-2DDE and GLCM-Entropy (i.e., Shannon Entropy) for the Port Scan attack. The blue plot shows the trend of the specific feature while the bar graph (purple bars superimposed on the plot) identifies the windows where the attack is implemented and labelled. It can be seen in Figure 5a that the values of GLCM-2DDE (called simply 2DDE in the rest of this paper) are notably higher in correspondence to the Port Scan attack than the normal legitimate traffic. This difference is less evident for the GLCM-Entropy feature. These differences in values are the reason why the performance of GLCM-2DDE is higher than the GLCM-Entropy when ML is applied.

We highlight that Figure 5 and the previous paragraph are only used for informational purposes to provide to the reader with a visual recognition of the difference of the trends in the data set once two different entropy measures are applied to the network flows data. Figure 5 is not used to select features for the classification phase because the SFS is used for this purpose as described in Section 4.

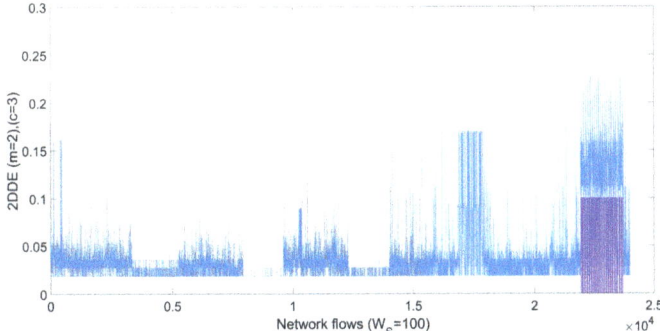

(a) Trend of the GLCM and 2DDE feature with $m = 2$ and $c = 3$ ($F_{ID} = 6$) on the CICIDS2017 data set (DDoS and legitimate traffic only).

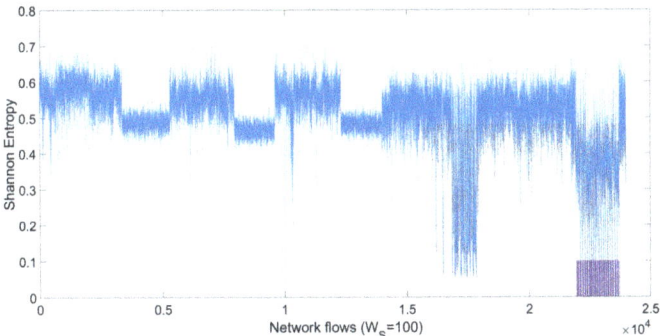

(b) Trend of the GLCM and Shannon Entropy feature ($F_{ID} = 5$) on the CICIDS2017 data set (DDoS and legitimate traffic only).

Figure 5. Trends of two features on the CICIDS2017 data set (DDoS and legitimate traffic only) and $W_S = 100$. The purple bars indicate the labels of the DDoS attack.

4. Results

4.1. Optimization

This sub-section provides the results on the optimization of the hyperparameters described in the previous sections.

A grid approach was used to determine the optimum values of the hyperparameters. While, other methods (e.g., gradient, meta-heuristics algorithms) could be more efficient, it should be considered that the ranges of values for each hyperparameter are quite limited. In addition, the intention is to show in an explicit way the impact of each hyperparameter for the detection performance. The metric is used to determine the optimal values of the hyperparameters.

The summary of the hyperparameters used in this study, the optimal values and the range of the hyperparameters are shown in Table 3. In the rest of this sub-section and related figures, we show how a specific hyperparameter impacts the detection accuracy of the threat both for DDoS attack and Port Scan attack. For each presented result, the other hyperparameters are set to the values identified in Table 3. The Decision Trees (DT) ML algorithm was used to generate the results provided in this sub-section. As shown in Section 4.3 the DT algorithm has a higher detection performance than the SVM and Naive Bayes algorithms.

Table 3. Summary of the hyperparameters in the proposed approach and related optimal values.

Hyper-Parameter	Description	Range	Optimal Value
Q_F	GLCM quantization function	[12,16,20,24,28, 32,36,40,44,48]	DDoS $Q_F = 44$, Port Scan $Q_F = 40$
W_S	Size of the sliding window	[100,200,300,400,500]	DDoS and Port Scan $W_S = 100$
G_D	GLCM distance	[1,2,3,4]	DDoS $G_D = 2$ and Port Scan $G_D = 1$
$N_B(DT)$	Number of branches in the Decision Tree algorithm	[4 ... 20]	DDoS $N_B = 12$, Port Scan $N_B = 8$
γ and C (SVM)	γ and C in the Support Vector Machine algorithm	$2^{[4...12]}, 2^{[4...12]}$	DDoS, Port Scan $\gamma = 2^7$ and $C = 2^8$ (SVM)

The following figures describe the results for the evaluation of the proposed approach for different values of the hyperparameters and for the different features used in the study. In most cases, the evaluation of a single hyperparameter is provided while the other hyperparameters are set to the values described in Table 3 unless otherwise noted.

Figure 6a,b show respectively for the Port Scan and the DDoS attacks, the impact of the GLCM distance G_D for different values of the window size W_S. These results are obtained using all the 64 features identified in Table 2. It can be noted that the optimal value of W_S is 100 network flows, as the ER increases with larger values of W_S. This may due to the reasons that the difference between legitimate traffic and the traffic related to the attack are more evident when the W_S is relatively small. On the other side, $W_S = 100$ is the lower limit of W_S to allow the GLCM to operate on a grayscale picture large enough to obtain meaningful values. Figure 6a shows that a value of $G_D = 1$ is optimal to detect the Port Scan attack, while Figure 6a shows that a value of $G_D = 2$ is optimal for the DDoS attack. These results seem to indicate that there is no need to use values of G_D larger than 2, which would also be more computing intensive.

Then, the impact of the quantization factor Q_F was evaluated. As described before, the quantization factor in the GLCM definition is an important factor in the application of GLCM. A large value of Q_F provides an higher granularity which can be beneficial in the application of the ML algorithm for the detection of the threat. On the other side, a large value of Q_F is more computing expensive for the calculation of the GLCM features and 2DDE as the resulting GLCM matrix are larger (the GLCM size is $Q_F * Q_F$). This is an important trade-off, which was investigated for each specific feature and for each attack.

Figure 7a,b shows the impact of the Q_F parameter on the detection accuracy respectively for the Port Scan and the DDoS attack for the first 8 features (only the first 8 features are provided in these figures for reasons of space, but subsequent figures will consider all features). The value of W_S is set to 100 since the previous Figure 6 has shown that $W_S = 100$ is the optimal value for attack detection. Figure 7a,b provide two important results: the first is that they identify the optimal value of the Q_F parameter ($Q_F = 40$ for the Port Scan attack and $Q_F = 44$ for the DDoS attack). The second is that they show that the 2DDE features have a better performance than the other features. This result justifies the assumption done in this paper for the application of 2DDE to the problem of IDS.

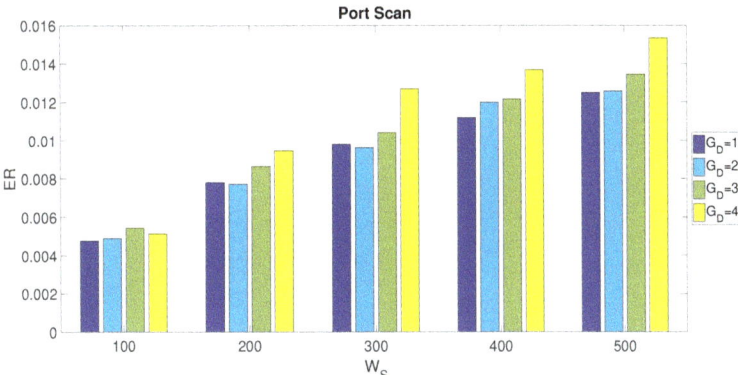

(**a**) Error Rate (ER) dependence on GLCM distance G_D for Port Scan attack.

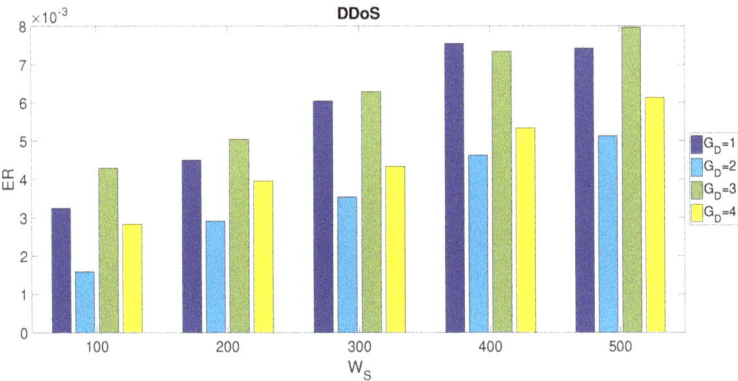

(**b**) Error Rate (ER) dependence on GLCM distance G_D for DDoS attack.

Figure 6. Dependence on GLCM distance G_D and Window size W_S using best selected features. DT algorithm is used.

Figure 7 shows only the first 8 features. Then, a more extensive analysis of the detection performance of each of the 64 features was carried on by setting the optimal value of the other hyperparameters (G_D, Q_F and W_S). The results are shown in Figure 8a,b where the ER is reported for each feature identified with the F_{ID} identifier. To better visualize the features related to 2DDE a red bar is used in the Figures. Figure 8a,b show that the 2DDE is able to obtain a consistent high detection accuracy in comparison to the other features for all the 64 features. In particular, for both attacks, the values of $m = 2$ and $c = 3$ in the 2DDE definition provides a better performance than the values of $m = 3$ and $c = 2$ in the 2DDE definition. This result shows the higher detection performance of 2DDE in comparison to the other features (e.g., Shannon entropy or variance). The results shown in these figures also give an indication on the GLCM angle, which is most performing. In general, the GLCM distance and angle defined by the 2-tuple [0 G_D] (which corresponds to $F_{ID} = 1 \ldots 8$) provides better results (in terms of detection accuracy) than the other 2-tuples.

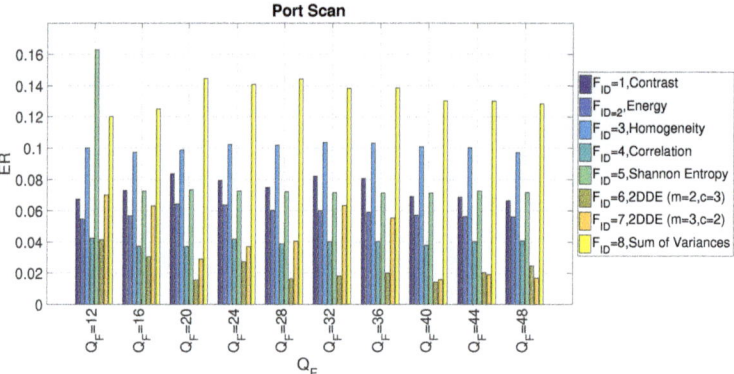

(**a**) Error Rate (ER) vs. quantization factor of the GLCM Q_F for the Port Scan attack with W_S=100 for the first 8 features ($F_{ID} = 1\ldots 8$).

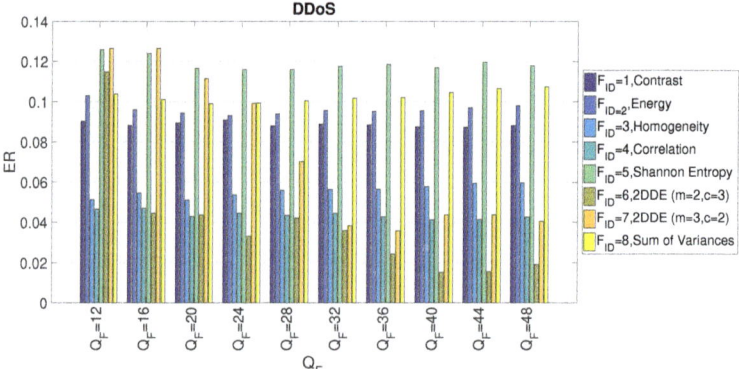

(**b**) Error Rate (ER) vs. quantization factor of the GLCM Q_F for the DDoS attack with $W_S = 100$ for the first 8 features ($F_{ID} = 1\ldots 8$).

Figure 7. Dependence on GLCM distance G_D and W_S using the first 8 features ($F_{ID} = 1\ldots 8$). DT algorithm is used.

The importance of 2DDE in comparison to other features for the IDS problem is also visible, once SFS is applied to select the optimal set of features on the basis of the value of hyperparameters already set. The results of the application of SFS is presented in Table 4, where the 10 best features are shown respectively for the DDoS and the Port Scan attack. In Table 4, the 2DDE features are highlighted in red. It can be seen that the 2DDE features are substantially present among the 10 best features, which shows the the application of 2DDE to this specific problem is an important element to achieve an higher detection accuracy of the attack.

Table 4. Ten best features obtained for the Port Scan and the DDoS attack using the SFS approach. DT algorithm is used.

Attack	Ten Best Features
Port Scan ($W_S = 100$, $G_D = 1$)	[2,7,13,15,38,42,49,53,54,62]
DDoS ($W_S = 100$, $G_D = 2$)	[5,15,24,32,34,54,56,59,63,64]

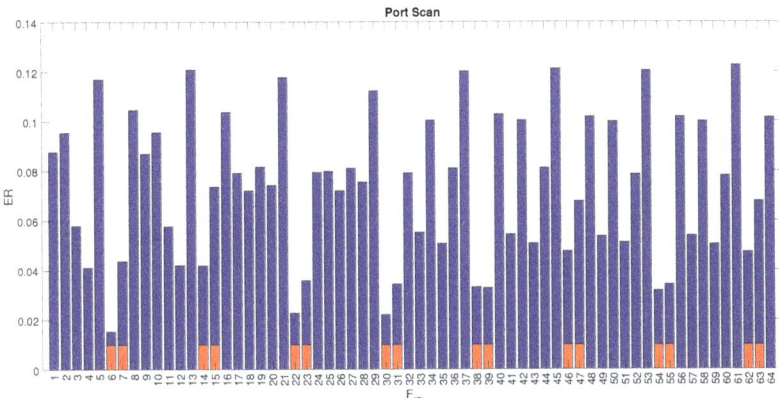

(a) Error Rate (ER) vs. all features for the PortScan attack with $W_S = 100$, $Q_F = 40$ and $G_D = 1$.

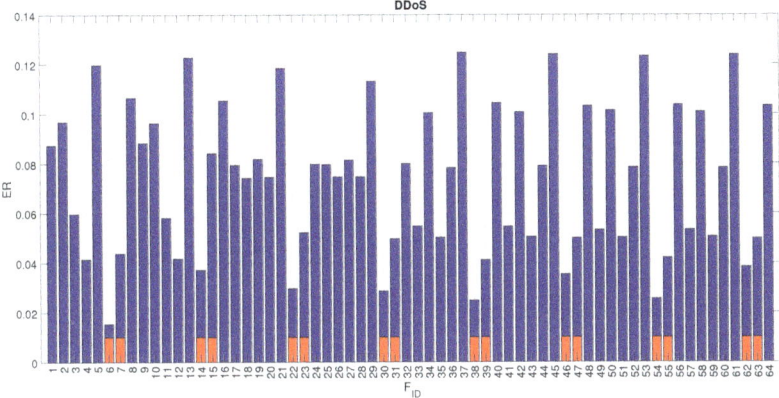

(b) Error Rate (ER) vs. all features for the DDoS attack with $W_S = 100$, $Q_F = 44$ and $G_D = 2$.

Figure 8. Error Rate (ER) relation with all features with $W_S = 100$. The features related to 2DDE are highlighted with a red bar for improved visualization. DT algorithm is used.

4.2. Optimized Results

On the basis of the best features described in Table 4 and the optimal values of the hyper-parameters defined in Table 3, the ER, FPR and FNR have been calculated using the Decision Tree algorithm. It was also evaluated the impact of the size of the data set. From the whole data set, a partitions of the whole data set have been selected and the ER, FPR and FNR have been calculated. The results are presented in Figure 9 and related subfigures where 'All' means the whole data set and 'All/x' is a partition by the factor x. The size of 'All' can be calculated from the values presented in Table 1. The partition is created by extracting randonmly 'All/x' elements from the whole data set. To mitigate the risk of bias, the selection of the partition and the calculation of the results is repeated 100 times and the results are averaged.

Both for the PortScan and the DDoS attacks, it can be seen that the performance of the detection of the attack is lower for smaller partitions of the data set because it is more difficult for the algorithm to discriminate the legitimate traffic from the traffic related to the attack. This trend is coherent for all the three metrics (ER, FPR and FNR) and the two attacks.

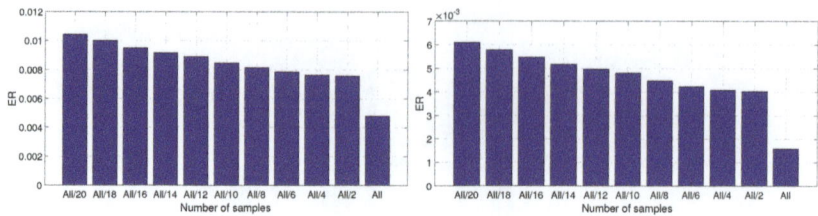

(a) ER for the PortScan attack for different sizes of the data set. 'All' means the whole data set.
(b) ER for the DDoS attack for different sizes of the data set. 'All' means the whole data set.

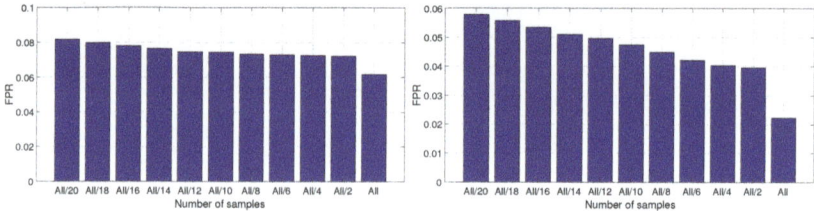

(c) FPR for the PortScan attack for different sizes of the data set. 'All' means the whole data set.
(d) FPR for the DDoS attack for different sizes of the data set. 'All' means the whole data set.

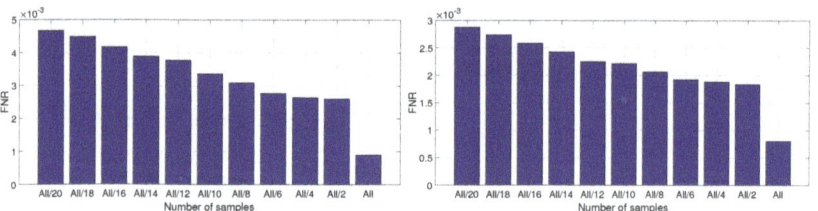

(e) FNR for the PortScan attack for different sizes of the data set. 'All' means the whole data set.
(f) FNR for the DDoS attack for different sizes of the data set. 'All' means the whole data set.

Figure 9. ER, FPR and FNR for the PortScan and DDoS attack for different sizes of the whole data set.

To complete the previous results, the ROCs for the DDoS and the PortScan attacks are presented respectively in Figure 10a,b. Since the FPR is relatively limited (because the data set is quite unbalanced), a more detailed figure of the same ROCs (i.e., zoom of the previous figures) is presented in Figure 10a,b respectively for the DDoS and the PortScan attacks. The values of the EER for each value of W_S are also reported. The results from the ROCs confirm the previous result that the optimal value of the window size is $W_S = 100$ because an increase of W_S produces slightly worst results in terms of ROCs and EER. It can also been seen that the detection of the PortScan attack is slightly worse than the DDoS attack. This may be due to the reason that PortScan attacks are more difficult to distinguish from legitimate traffic than the DDoS attacks when the entropy measures are applied (especially in the CIC-IDS2017 data set). The structure of the sequences of network flows features in the DDoS attacks can be quite different from legitimate traffic (e.g., since a flooding of messages is implemented) while the PortScan attack traffic may resemble legitimate traffic. The weakness of the proposed approach in achieving an optimal FPR is also discussed in the comparison with the literature results in Section 4.3. We note that the proposed approach manages to achieve a very competitive FNR instead.

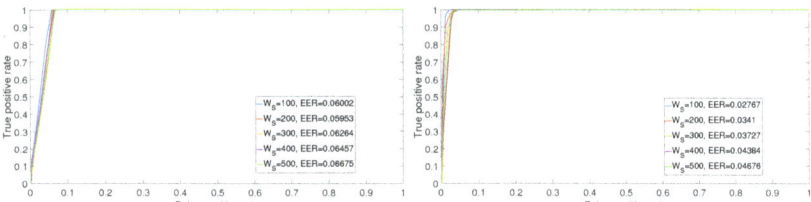

(**a**) ROCs and related EERs for the PortScan attack. (**b**) ROCs and related EERs for the DDoS attack.

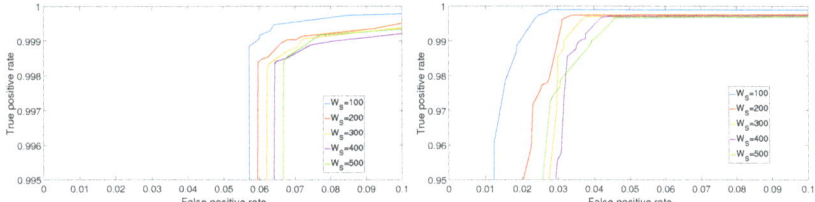

(**c**) Detailed view of the ROC for the Port Scan attack for different values of W_S.
(**d**) Detailed view of the ROC for the DDoS attack for different values of W_S.

Figure 10. ROCs and related EERs for the PortScan and DDoS attack for different values of W_S. DT with optimal hyperparameter values from Table 3 and optimal set of features from Table 4. The bottom figures show the detailed view of the ROCs.

4.3. Comparison with Other Studies

On the basis of the optimization results obtained in the previous Section 4.1, we have calculated the values of ER, FPR and FNR for the Port Scan and the DDoS attack and we compared these results with the results in literature on the same CICIDS2017 data set. The comparison is indicative because each study may have modified the initial data set in different ways: a subset of the initial 78 features may be used or the data pertaining only to specific attacks has been used. We must also consider that the CICIDS2017 data set is relatively recent and not all the studies using it focused on a specific attack as it was done in this study. The results are presented in Table 5 where the first three columns identify the value of ER, FPR and FNR. The fourth column provides relevant notes (e.g., the specific adopted algorithm). The fifth column identifies the specific attack (i.e., DDoS or Port Scan) and the related study where the results were produced. Table 5 does also provides the comparison of the machine learning algorithms: SVM algorithm, Naive Bayes algorithm and Decision Tree.

The results show that the proposed approach is competitive against other approaches proposed in literature. For example, in the case of the DDoS attack, the obtained ER (0.0016) is smaller than the ER obtained by most of the other results with the exception of the study [7] where it has the same value or the study [6] where the obtained ER is slightly lower than the result obtained in this study (0.0015 rather than 0.0016). It has to be noted that both [6,7] use sophisticated DL algorithms which are more computing demanding than the approach proposed in this paper. In addition, it is noted that the approach proposed in this paper is able to obtain a value of False Negative Rate (FNR) for the DDoS attack (i.e., 0.00079), which is considerable lower than the result obtained by all other approaches. On the other side, the FPR is worse than the value obtained by the other studies. Then, this approach is particularly strong on the FNR performance but it is weaker on the FPR. A potential reason why FNR is so low in comparison to literature is due to the sliding window approach where the presence of only a single network flow labelled as an attack in the data set is magnified to the size of the sliding window. The improvement of the FPR is one of the actions for future developments and investigations on this approach.

Table 5. Summary table of the ER, FPR and FNR results obtained with this approach (different machine learning algorithms) and the results from literature.

ER	FPR	FNR	Optimal Values and/Or Notes	Approach-Attack
0.0016	0.0223	0.0008	$N_B = 12$	This approach (Decision Tree), DDoS
0.0084	0.0354	0.0069	$\gamma = 2^7, C = 2^{10}$	This approach (SVM), DDoS
0.0248	0.02	0.0251	none	This approach (Naive Bayes), DDoS
0.0045	0.0023	0.0476	Online Kernel Online Anomaly Detection (KOAD)	[10], DDoS
0.0016	N/A	0.0016	Deep Belief Network (DBN) and Bidirectional Gated Recurrent Unit (BiGRU)	[7] DDoS
0.0015	0.0017	0.0068	Deep Neural Networks (DNN)	[6], DDoS
0.0048	0.0618	0.0009	$N_B = 17$	This approach (Decision Tree), Port Scan
0.0082	0.0653	0.0038	$\gamma = 2^8, C = 2^9$	This approach (SVM), Port Scan
0.0339	0.0571	0.0322	none	This approach (Naive Bayes), Port Scan
N/A	0.0094	0.0078	cost-sensitive differential evolution classifier	[8], Port Scan
0.0051	0.004	0.0016	LSTM	[9], Port Scan

The results obtained with the DDoS attack are confirmed by the results obtained by the PortScan attack. The obtained FNR is better than the results obtained in literature while the ER is also smaller than the results presented in other studies. In particular, our approach achieves a similar ER to the results in [9], which uses a DL approach (i.e., LSTM). On the other side, the FPR obtained with this approach is higher than the results obtained in literature. Another result shown in Table 5 is that the Decision Tree algorithm has a better detection performance than the SVM and Naive Bayes algorithms. This result is consistent with [5] where the DT provided the optimal detection accuracy.

An evaluation of the use of all the GLCM angles was also implemented to validate the adoption of only a limited set of GLCM angles as described in Section 3.3. The results are provided in Table 6 using the Decision Tree algorithm. The results in Table 6 show that a subset of the GLCM angles (as selected in this study) provides a better performance than using all angles since the ERs for the subset are smaller than the ERs for all the GLCM angles. The results are consistent for different values of W_S and for both attacks of PortScan and DDoS.

4.4. Computing Times

In the following Table 7, we report the computing time of the proposed approach with the application of ML directly on the data set in a similar way to what was done in the paper [5]. The approach proposed in this paper implements a dimensionality reduction and the computing time to execute the machine learning algorithm on the reduced set is minimal. On the other side, the time requested to calculate the GLCM is significant (34 s in average for the DDoS attack and 31 s in average for the PorScan attack) as shown in Table 7. The average time needed to calculate the 2DDE entropy measure is also relatively high (63 s for the DDoS attack and 83 s for the PortScan attack). These calculated times are based on $W_S = 100$ since this was the window size with the minimum ER and the optimal selection of features presented in Table 4. In this study, it was used a laptop with Intel i7 85550U CPU running at 1.8 GHz with 16 GBytes of RAM and no GPU.

Table 6. Comparison on the set of GLCM angles using Error Rate (ER): subset of angles used in this study in comparison to the use of all the GLCM angles.

Attack and Set of Angles	$W_S = 100$	$W_S = 200$	$W_S = 300$	$W_S = 400$	$W_S = 500$
Port Scan (all angles: 128 features)	0.0081	0.0279	0.0311	0.0311	0.0292
Port Scan (subset of angles: 64 features)	0.0048	0.0078	0.0098	0.0112	0.0125
DDoS (all angles: 128 features)	0.0026	0.0032	0.0045	0.0048	0.0062
DDoS (subset of angles: 64 features)	0.0016	0.0029	0.0035	0.0046	0.0051

Table 7. Computing times in seconds (s).

Approach Step	Computing Time (s)	Attack
Decision Tree algorithm execution	1 s	DDoS
GLCM computation	34 s GLCM	DDoS
2DDE computation	63 s GLCM	DDoS
Decision Tree algorithm execution	1 s	PortScan
GLCM computation	31 s GLCM	PortScan
2DDE computation	83 s GLCM	PortScan

5. Conclusions

This study proposes a novel approach for IDS based on anomaly detection which is based on the transformation of the network flows metrics into grayscale images. Then, the Gray-Level Co-occurrence Matrices (GLCM) are calculated on the grayscale images and features are calculated on the GLCM. Beyond the application of well known GLCM Haralick features (i.e., contract, homogeneity, entropy), this paper proposes the novel application of 2D Dispersion Entropy (2DDE) recently proposed in literature. The results show that the application of 2DDE to GLCM significantly enhances the detection accuracy of the proposed IDS. The approach is applied to the recently published CICIDS2017 data set for two specific attacks: DDoS and Port Scan. The results of this approach are compared with the results obtained by other studies on the same CICIDS2017 data set obtaining an Error Rate (ER) which is higher or comparable with the results obtained with more sophisticated approach based on Deep Learning, which requires considerable more computing resources than our proposed approach. In addition, the False Negative Rate (FNR) obtained with our approach is significantly better than all the other results obtained in literature. On the other side, the False Positive Rate (FPR) is slightly worse than the results obtained in literature. This may due to the possibility that the transformation of the network flows features to gray level images and then GLCM-base features has the tendency to lose the specific characteristics of the attack related traffic in comparison to the normal traffic.

Future developments will try to improve the FPR by adopting improvements of the proposed approach in different directions. One direction would be to use Fuzzy Gray-Level Co-occurrence Matrices since it has demonstrated a superior performance in some applications, but it has not been used in IDS problems. Another direction would be the application of non linear GLCM where the quantization factor is calculated in a non linear way. The significant number of hyperparameters to tune in the approach (both in the GLCM definition and 2D dispersion entropy definition) is also a challenge to mitigate for a practical deployment of this approach. One possibility to resolve the challenge would be to investigate the application of meta-heuristics algorithms (e.g., particle swarm optimization) to automatically tune the hyperparameters. Another possibility would be to investigate the hyperparameters optimization in other data sets to generalize the selection of the optimal values. Finally, the combination of GLCM together with Deep Learning algorithms will also be considered. For example, Convolutional Neural Networks (CNN) could be applied to the GLCM representations rather than the initial gray-scale images derived directly from the network flows statistics.

Author Contributions: Conceptualization, G.B., J.L.H.R. and I.A.; methodology, G.B.; writing G.B., J.L.H.R., I.A.; funding acquisition, G.B., J.L.H.R. All authors have read and agreed to the published version of the manuscript

Funding: This work has been partially supported by the European Commission through project SerIoT funded by the European Union H2020 Programme under Grant Agreement No. 780139. The opinions expressed in this paper are those of the authors and do not necessarily reflect the views of the European Commission.

Institutional Review Board Statement: Not applicable.

Informed Consent Statement: Not applicable.

Data Availability Statement: This study used the public data set described in [5].

Acknowledgments: We are thankful to Anne Humeau-Heurtier and the other authors of [4] to graciously provide the MATLAB code for the implementation of the 2D Dispersion Entropy.

Conflicts of Interest: The authors declare no conflict of interest.

Abbreviations

The following abbreviations are used in this manuscript:

CNN	Convolutional Neural Networks
2DDE	2D Dispersion Entropy
DDoS	Distributed Denial of Service
DL	Deep Learning
DT	Decision Tree
ER	Error Rate
EER	Equal Error Rate
FP	False Positives
FPR	False Positives Rate
FN	False Negatives
FNR	False Negatives Rate
GLCM	Gray-Level Co-occurrence Matrices
KOAD	Kernel Online Anomaly Detection
NCDF	Normal Cumulative Distribution Function
IDS	Intrusion Detection Systems
ML	Machine Learning
NCDF	Normal Cumulative Distribution Function
RBF	Radial Basis Function
ROC	Receiver Operating Characteristics
SFS	Sequential Feature Selection
SVM	Support Vector Machine

References

1. Lunt, T.F. A survey of intrusion detection techniques. *Comput. Secur.* **1993**, *12*, 405–418. [CrossRef]
2. Liao, H.J.; Lin, C.H.R.; Lin, Y.C.; Tung, K.Y. Intrusion detection system: A comprehensive review. *J. Netw. Comput. Appl.* **2013**, *36*, 16–24. [CrossRef]
3. Dromard, J.; Roudière, G.; Owezarski, P. Online and scalable unsupervised network anomaly detection method. *IEEE Trans. Netw. Serv. Manag.* **2016**, *14*, 34–47. [CrossRef]
4. Azami, H.; da Silva, L.E.V.; Omoto, A.C.M.; Humeau-Heurtier, A. Two-dimensional dispersion entropy: An information-theoretic method for irregularity analysis of images. *Signal Process. Image Commun.* **2019**, *75*, 178–187. [CrossRef]
5. Sharafaldin, I.; Lashkari, A.H.; Ghorbani, A.A. *Toward Generating a New Intrusion Detection Dataset and Intrusion Traffic Characterization*; ICISSP: Funchal, Portugal, 2018; pp. 108–116.
6. de Souza, C.A.; Westphall, C.B.; Machado, R.B.; Sobral, J.B.M.; dos Santos Vieira, G. Hybrid approach to intrusion detection in fog-based IoT environments. *Comput. Netw.* **2020**, *180*, 107417. [CrossRef]
7. Yu, X.; Li, T.; Hu, A. Time-series Network Anomaly Detection Based on Behaviour Characteristics. In Proceedings of the 2020 IEEE 6th International Conference on Computer and Communications (ICCC), Chengdu, China, 11–14 December 2020; pp. 568–572.
8. Al-Sawwa, J.; Ludwig, S.A. Performance evaluation of a cost-sensitive differential evolution classifier using spark–Imbalanced binary classification. *J. Comput. Sci.* **2020**, *40*, 101065. [CrossRef]
9. Hossain, M.D.; Ochiai, H.; Fall, D.; Kadobayashi, Y. LSTM-based Network Attack Detection: Performance Comparison by Hyper-parameter Values Tuning. In Proceedings of the 2020 7th IEEE International Conference on Cyber Security and Cloud Computing (CSCloud)/2020 6th IEEE International Conference on Edge Computing and Scalable Cloud (EdgeCom), New York, NY, USA, 1–3 August 2020; pp. 62–69.
10. Çakmakçı, S.D.; Kemmerich, T.; Ahmed, T.; Baykal, N. Online DDoS attack detection using Mahalanobis distance and Kernel-based learning algorithm. *J. Netw. Comput. Appl.* **2020**, *168*, 102756. [CrossRef]
11. Moustafa, N.; Hu, J.; Slay, J. A holistic review of network anomaly detection systems: A comprehensive survey. *J. Netw. Comput. Appl.* **2019**, *128*, 33–55. [CrossRef]
12. Zarpelão, B.B.; Miani, R.S.; Kawakani, C.T.; de Alvarenga, S.C. A survey of intrusion detection in Internet of Things. *J. Netw. Comput. Appl.* **2017**, *84*, 25–37. [CrossRef]
13. Liu, H.; Lang, B. Machine Learning and Deep Learning Methods for Intrusion Detection Systems: A Survey. *Appl. Sci.* **2019**, *9*, 4396. [CrossRef]
14. Sommer, R.; Paxson, V. Outside the closed world: On using machine learning for network intrusion detection. In Proceedings of the 2010 IEEE Symposium on Security and Privacy, Berleley/Oakland, CA, USA, 16–19 May 2010; pp. 305–316.
15. Behal, S.; Kumar, K.; Sachdeva, M. D-FACE: An anomaly based distributed approach for early detection of DDoS attacks and flash events. *J. Netw. Comput. Appl.* **2018**, *111*, 49–63. [CrossRef]
16. Radivilova, T.; Kirichenko, L.; Alghawli, A.S. Entropy Analysis Method for Attacks Detection. In Proceedings of the 2019 IEEE International Scientific-Practical Conference Problems of Infocommunications, Science and Technology (PIC S&T), Kiev, Ukraine, 8–11 October 2019; pp. 443–446.
17. Shah, S.B.I.; Anbar, M.; Al-Ani, A.; Al-Ani, A.K. Hybridizing entropy based mechanism with adaptive threshold algorithm to detect ra flooding attack in ipv6 networks. In *Computational Science and Technology*; Springer: Berlin/Heidelberg, Germany, 2019; pp. 315–323.
18. Zhang, Y.; Chen, X.; Jin, L.; Wang, X.; Guo, D. Network intrusion detection: Based on deep hierarchical network and original flow data. *IEEE Access* **2019**, *7*, 37004–37016. [CrossRef]
19. Lopez-Martin, M.; Carro, B.; Sanchez-Esguevillas, A.; Lloret, J. Conditional variational autoencoder for prediction and feature recovery applied to intrusion detection in iot. *Sensors* **2017**, *17*, 1967. [CrossRef]
20. Zhou, H.; Wang, Y.; Lei, X.; Liu, Y. A method of improved CNN traffic classification. In Proceedings of the 2017 13th International Conference on Computational Intelligence and Security (CIS), Hong Kong, China, 15–18 December 2017; pp. 177–181.
21. McHugh, J. Testing intrusion detection systems: a critique of the 1998 and 1999 darpa intrusion detection system evaluations as performed by lincoln laboratory. *ACM Trans. Inf. Syst. Secur. (TISSEC)* **2000**, *3*, 262–294. [CrossRef]
22. Lopez-Martin, M.; Carro, B.; Sanchez-Esguevillas, A. Variational data generative model for intrusion detection. *Knowl. Inf. Syst.* **2019**, *60*, 569–590. [CrossRef]
23. Abdulhammed, R.; Musafer, H.; Alessa, A.; Faezipour, M.; Abuzneid, A. Features dimensionality reduction approaches for machine learning based network intrusion detection. *Electronics* **2019**, *8*, 322. [CrossRef]
24. Vijayanand, R.; Devaraj, D. A novel feature selection method using whale optimization algorithm and genetic operators for intrusion detection system in wireless mesh network. *IEEE Access* **2020**, *8*, 56847–56854. [CrossRef]
25. Maseer, Z.K.; Yusof, R.; Bahaman, N.; Mostafa, S.A.; Foozy, C.F.M. Benchmarking of machine learning for anomaly based intrusion detection systems in the CICIDS2017 dataset. *IEEE Access* **2021**, *9*, 22351–22370. [CrossRef]
26. Baldini, G.; Giuliani, R.; Steri, G.; Neisse, R. Physical layer authentication of Internet of Things wireless devices through permutation and dispersion entropy. In Proceedings of the 2017 Global Internet of Things Summit (GIoTS), Geneva, Switzerland, 6–9 June 2017; pp. 1–6.

27. Rostaghi, M.; Azami, H. Dispersion entropy: A measure for time-series analysis. *IEEE Signal Process. Lett.* **2016**, *23*, 610–614. [CrossRef]
28. Shawe-Taylor, J.; Cristianini, N. *Support Vector Machines*; Cambridge University Press: Cambridge, UK, 2000; Volume 2.
29. Rish, I. An empirical study of the naive Bayes classifier. In Proceedings of the IJCAI 2001 Workshop Empirical Methods in Artificial Intelligence, Seattle, WA, USA, 4–6 August 2001; pp. 41–46.
30. Haralick, R.M.; Shanmugam, K.; Dinstein, I.H. Textural features for image classification. *IEEE Trans. Syst. Man Cybern.* **1973**, *SMC-3*, 610–621. [CrossRef]
31. Haralick, R.M. Statistical and structural approaches to texture. *Proc. IEEE* **1979**, *67*, 786–804. [CrossRef]

Article

New Subclass Framework and Concrete Examples of Strongly Asymmetric Public Key Agreement

Satoshi Iriyama [1,†], Koki Jimbo [1,†,*] and Massimo Regoli [2,†]

[1] Information Science Department, Tokyo University of Science, 2641, Yamazaki, Noda, Chiba 278-8510, Japan; iriyama@is.noda.tus.ac.jp

[2] DICII, Engineering Faculty Via del Politecnico, Universitá di Roma Tor Vergata, 1, 00133 Roma, Italy; regoli@uniroma2.it

* Correspondence: 6319702@ed.tus.ac.jp

† These authors contributed equally to this work.

Abstract: Strongly asymmetric public key agreement (SAPKA) is a class of key exchange between Alice and Bob that was introduced in 2011. The greatest difference from the standard PKA algorithms is that Bob constructs multiple public keys and Alice uses one of these to calculate her public key and her secret shared key. Therefore, the number of public keys and calculation rules for each key differ for each user. Although algorithms with high security and computational efficiency exist in this class, the relation between the parameters of SAPKA and its security and computational efficiency has not yet been fully clarified. Therefore, our main objective in this study was to classify the SAPKA algorithms according to their properties. By attempting algorithm attacks, we found that certain parameters are more strongly related to the security. On this basis, we constructed concrete algorithms and a new subclass of SAPKA, in which the responsibility of maintaining security is significantly more associated with the secret parameters of Bob than those of Alice. Moreover, we demonstrate 1. insufficient but necessary conditions for this subclass, 2. inclusion relations between the subclasses of SAPKA, and 3. concrete examples of this sub-class with reports of implementational experiments.

Keywords: public key exchange; security; asymmetric; asymmetric algorithm; cryptography; framework; limited computational power; computationally biased

1. Introduction

Since Shannon proposed the concept of "perfect secrecy" in crypto-systems [1], in which he introduced a theoretically unbreakable system even against computational power, the distribution of secret keys between the sender (Alice) and receiver (Bob) via an insecure channel has been one of the greatest problems in cryptography.

The Diffie–Hellman (DH) public key agreement (PKA) protocol that was proposed in 1976 [2] and the Rivest–Shamir–Adleman (RSA) crypto-system that was presented in 1978 [3] represented the most significant works in the area of cryptography, and it was previously believed that the problem of key distribution had been resolved.

However, recent considerable developments in the computational power of eavesdroppers have introduced several potential (even if not immediate) threats against standard PKA algorithms and public key cryptographies, particularly for small key lengths [4]. To maintain security, users have been forced to select longer keys, and the increased key length has led to higher computational costs. Thus, the preparation of secure communication infrastructure, particularly for devices with limited memory and computational power, has become challenging. Furthermore, the threat of quantum computers that are currently under development and Shor's algorithm [5] cannot be underestimated.

Considering the demand for algorithms that are resilient against any type of theoretical attack, including quantum algorithm-based attacks, the development and study of new PKA algorithms and public key cryptographies, namely post-quantum cryptography

(PQC), has become widespread. PKAs and public key cryptographies based on lattice problems such as the shortest vector problem (SVP), closest vector problem, and learning with errors (LWE) are among the most well known methods. Among these, SVP-based PKA and public key cryptographies, including NTRU prime [6], NTRU-HRSS-KEM [7], and NTRU Encrypt [8], module LWE-based PKA such as CRYSTALS–Kyber [9], and ring LWE-based PKA including NewHope [10] are leading approaches in this research area and have been considered as candidates for the NIST (National Institute of Standards and Technology) standardization of PQC systems [11,12]. When the parameters are properly selected, the above algorithms are considered to be resilient against attacks that use quantum computers and sufficiently computationally efficient to be used in practice.

However, there has been substantial discussion regarding the security of such algorithms. For example, in certain LWE-based algorithms, even if sufficiently large parameters are selected, a possible weakness has been observed [13–15]. Moreover, the notion that the difficulty in solving ring LWE is equivalent to that of solving the LWE (the difficulty of LWE is discussed in [14,16]) has not yet been proven; thus, other cases of weakened security [9,10] may arise. Weak parameters for NTRU-type PKA algorithms and public key cryptography are also reported in [17]. Owing to these uncertainties in the parameter settings to maintain security even in an ideal situation (i.e., without assuming limited memory and computational power), the preparation of secure communication infrastructure with these new-generation algorithms for less capable devices has resulted in greater difficulty and insecurity. Although there is no doubt that these algorithms will offer significant benefits even after the post-quantum computer era, security analysis of these algorithms should continue until users can be provided with a "guide" that explains how to set parameters to ensure secure PKA and public key cryptography according to the needs and environments of users.

1.1. Research Concept and Goals

We define a function $C : \mathbb{N} \times \mathbb{R}^+ \to \mathbb{R}^+$, which shows the computational costs for T calculation steps by a device with efficiency E_D (calculation steps per time) as follows:

$$C(T, E_D) := T \times \frac{1}{E_D}.$$

Thus, it is given by time (s, ms, or another unit). Let PKA be a set of PKA algorithms and let $T_A : PKA \times \mathbb{N} \to \mathbb{N}$ be a function that shows the calculation steps required for Alice to calculate the $N \in \mathbb{N}$ bit length of the secret shared key (SSK) of an algorithm $Alg \in PKA$, which increases monotonically for N. T_B denotes the calculation steps for Bob in a similar manner.

Next, we consider $Alg \in PKA$, which has the following relation:

$$T_A(Alg, N) = T_B(Alg, N) \tag{1}$$

for all $N \in \mathbb{N}$. In this case, we suppose that the maximal computational cost that Bob is allowed to incur for the SSK calculation of $Alg \in PKA$, denoted by C_{max}^B, is $C_{max}^B = T_B(Alg, N_0) \times \frac{1}{E_B}$, which is achieved when the bit length of the SSK is some $N_0 \in \mathbb{N}$, and Bob can compute their SSK for all $N \in \mathbb{N}$ to satisfy:

$$C_B(T_B(Alg, N), E_B) := T_B(Alg, N) \times \frac{1}{E_B} \leq C_{max}^B := T_B(Alg, N_0) \times \frac{1}{E_B}, \tag{2}$$

where E_B denotes the device efficiency of Bob. As T_B is a monotonically increasing function for the bit size of the SSK, condition (2) is reduced to $N \leq N_0$. Furthermore, if $E_B = E_A$, where E_A denotes the device efficiency of Alice, the SSK can be computed for all $N \leq N_0$. Thus, Alice and Bob can calculate the SSK of a bit size that is equal to or less than N_0 within time C_{max}^B.

For the same $Alg \in PKA$, we assume that $E_B > E_A$ and the maximal computational cost that Bob is allowed to incur C_{max}^B is the same as (2), where N_0 is the smallest bit size of SSK to maintain security. Let Alice's computational cost for calculating her SSK of $N \in \mathbb{N}$ bits be $C_A(T_A(Alg, N), E_A) := T_A(Alg, N) \times \frac{1}{E_A}$, and if Alice needs to calculate her SSK within the cost C_{max}^B as well as Bob, the following relation must be satisfied:

$$C_A(T_A(Alg, N), E_A) = T_A(Alg, N) \times \frac{1}{E_A} \leq C_{max}^B = T_B(Alg, N_0) \times \frac{1}{E_B},$$

where the equality holds for some $N_1 \in \mathbb{N}$, but in this case, $N_1 < N_0$ must be satisfied because T_A is a monotonically increasing function for the bit size of the SSK and $\frac{1}{E_B} < \frac{1}{E_A}$. This observation indicates that they must either use an N_1-bit SSK, which is obviously less secure than when using an N_0-bit SSK, or let Alice incur a cost of $T_A(Alg, N_0) \times \frac{1}{E_A}$, which is larger than C_{max}^B. Most PKA algorithms, including the DH algorithm, satisfy (1), and there are many cases in which $E_B > E_A$ in modern society where IoT techniques are continually being developed; thus, this situation is inevitable in the near future, if not immediate.

We consider determining an algorithm denoted by $Alg_{A<B} \in PKA$, where

$$T_A(Alg_{A<B}, N) < T_B(Alg_{A<B}, N) \tag{3}$$

being satisfied for all $N \in \mathbb{N}$ is one solution to the above undesirable situation. We denote the maximal computational cost that Bob is allowed to incur as $C_{max,A<B}^B$, which is defined as $C_{max,A<B}^B := T_B(Alg_{A<B}, N_0) \times \frac{1}{E_B}$, where $N_0 \in \mathbb{N}$ and it is the smallest bit size of SSK to maintain security. In addition to the above $T_A = T_B$ case, we suppose that both Bob and Alice must calculate her SSK within the maximal computational cost that Bob is allowed to incur $C_{max,A<B}^B$. In this case, Alice can calculate all N bits of the SSK to satisfy

$$C_A(T_A(Alg_{A<B}, N), E_A) = T_A(Alg_{A<B}, N) \times \frac{1}{E_A} \leq C_{max,A<B}^B = T_B(Alg_{A<B}, N_0) \times \frac{1}{E_B}.$$

The equality holds when $T_A(Alg_{A<B}, N_1) \times \frac{1}{E_A} = T_B(Alg_{A<B}, N_0) \times \frac{1}{E_B}$ holds for some $N_1 \in \mathbb{N}$. In this case, it should be noted that $N_1 = N_0$ is achieved; that is, Alice calculating the SSK of N_0 bits within time $C_{max,A<B}^B$ is possible, provided that

$$\frac{T_A(Alg_{A<B}, N_0)}{T_B(Alg_{A<B}, N_0)} \leq \frac{E_A}{E_B} \tag{4}$$

holds, which is impossible when (1), because in this case, the left-hand side is equal to 1, but the right-hand side is less than 1. As $\frac{E_A}{E_B}$ is given (we may say that $\frac{E_A}{E_B}$ is a communication environment in which the algorithm is used), (4) is not always achieved for some $Alg_{A<B} \in PKA$ and $N_0 \in \mathbb{N}$. Conversely, we can determine the minimal environment $\frac{E_A}{E_B}$ where Alice and Bob can calculate N_0 bits SSK within $C_{max,A<B}^B$ time using $Alg_{A<B} \in PKA$ by simply calculating the left-hand side of (4).

Based on the above considerations, our research goals are as follows:

1. Constructions of $Alg_{A<B} \in PKA$.
2. For $Alg_{A<B} \in PKA$, the determination of $T_A(Alg_{A<B}, N_0')$ and $T_B(Alg_{A<B}, N_0')$ for any $N_0' \in \mathbb{N}$.
3. The construction of the PKA class to which $Alg_{A<B} \in PKA$ belongs and the introduction of conditions for PKA algorithms to be members of this class.

As mentioned above, goal 2 provides a lower bound of $\frac{E_A}{E_B}$ to calculate the SSK of N_0 bits within time $C_{max,A<B}^B$. Goal 3 provides instructions on how to construct PKA algorithms to possess the relation (3). Thus, improving algorithms such as those of [9,10] to possess this property may be possible by attempting to fix their parameters according to

the class conditions. We do not attempt to improve these algorithms in this study, but this subject is worthy of consideration and will be one of our most important future works.

We consider that these goals are achievable by fully utilizing the characteristics of the PKA framework known as strongly asymmetric public key agreement (SAPKA) [18]. The characteristics, high level of generality, and asymmetry of the key agreement process of SAPKA are explained in Section 1.2, along with its definition, and concrete methods that are derived from the characteristics are explained in Section 2.

Note that this study is not focused on how to construct secure PKA algorithms against any types of theoretical attacks; rather, it investigates how to reduce Alice's computational complexity while maintaining the security of one given PKA algorithm. Our main theorems (in Section 5) do not provide any instructions on how to enhance the security of PKA algorithms, and resilience against attack such as man-in-the-middle (MITM) attack is not discussed in this paper (we consider that these topics should be discussed after existence of $Alg_{A<B}$ is proven and mentioned in Section 7).

1.2. SAPKA Framework

We provide a brief definition of SAPKA (the explicit definition is presented in Section 2.4) and its characteristics in this section.

First, Bob prepares a multiplicative semi-group $\mathcal{S}$ with 1. Subsequently, he selects five maps:

$$x_1, x_2, x_3, x_4, N_1 : \mathcal{S} \to \mathcal{S},$$

where N_1 must be an easily invertible map. In this case, "easily" means that the calculation of $N_1^{-1} \circ N_1(y)$ for all $y \in \mathcal{S}$ can be performed in polynomial time. Furthermore, x_1, x_2, x_3, and x_4 must satisfy the following equation, which is known as the compatibility condition:

$$x_1 \circ x_2(y) = x_3 \circ x_4(y) \tag{5}$$

for all $y \in \mathcal{S}$, where $\circ$ denotes the map composition. Equation (5) is a condition for Alice and Bob to calculate the same SSK (see the key agreement process in Figure 1). The key agreement process of SAPKA can be described as in Figure 1, and every secret/public key is displayed in Table 1.

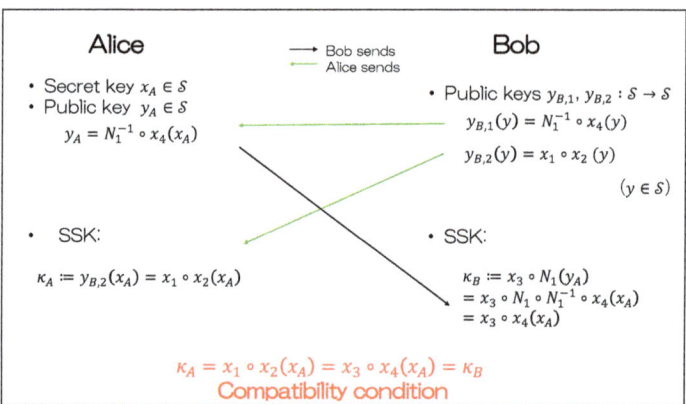

Figure 1. Key agreement process of SAPKA.

Table 1. Complete list of key of SAPKA.

	Key	Parameter
Bob	Secret keys	$x_1, x_2, x_3, x_4, N_1 : \mathcal{S} \to \mathcal{S}$
	Public keys	$y_{B,1} = N_1^{-1} \circ x_4 : \mathcal{S} \to \mathcal{S}$ $y_{B,2} = x_1 \circ x_2 : \mathcal{S} \to \mathcal{S}$
	SSK	$\kappa_B = x_3 \circ x_4(x_A) \in \mathcal{S}$
Alice	Secret key	$x_A \in \mathcal{S}$
	Public key	$y_A = N_1^{-1} \circ x_4(x_A) \in \mathcal{S}$
	SSK	$\kappa_A = x_1 \circ x_2(x_A) \in \mathcal{S}$

As can be observed from Figure 1 and Table 1, Bob's public keys are described by the map compositions and not by the element of $\mathcal{S}$. Sending a map means sending the calculation rule of the map in combination with a set of parameters, which is the domain of the map. Thus, Alice simply follows the rules of $y_{B,1}$ and $y_{B,2}$ to calculate y_A and κ_A, and to calculate these, she must first receive $y_{B,1}$ and $y_{B,2}$ from Bob. Regardless of the x_A that Alice selects (provided that $x_A \in \mathcal{S}$), the equality of κ_A and κ_B holds, because the compatibility condition (5) holds for all elements of $\mathcal{S}$. The generality mentioned above arises from the fact that there are only several restrictions for the secret keys of Bob, namely x_1, x_2, x_3, x_4, N_1, and semi-group $\mathcal{S}$. As the restrictions are only those in (5) and invertible regarding N_1, Bob has substantial freedom in terms of the choices of these maps and the algebraic structure. By fixing these maps and $\mathcal{S}$ concretely, various PKA algorithms can be described, including the most well known of these, namely the DH algorithm (presented in [18]). In this study, we do not attempt to describe new-generation algorithms such as [7,9,10] in the form of SAPKA. However, we are optimistic that these can be described because $\mathcal{S}$ can be selected as not only scalars but also matrices, for example, with numerous options for x_1, x_2, x_3, x_4, N_1.

Another notable characteristic of SAPKA is the asymmetry of the key agreement process. In this case, the asymmetry means that the number of public keys calculated by Alice and Bob differ, and thus, the two perform essentially different operations. Owing to this characteristic, an eavesdropper (Eve) must attempt attacks against a maximum of two public keys to obtain the secret information of either Bob or Alice. This may allow Alice to select her secret key from a set of small bit sizes and to reduce her computational complexity in certain cases. In Section 2, we explain Eve's strategies for recovering the SSK from public keys, an observation from her strategies, and the research method derived from this observation.

2. Methods and Abstract of This Study

We explain the strategies for Eve to recover the SSK and the method for our research goal, which can be derived from the observation of her strategies.

2.1. Eve's Strategies for Recovering SSK

Eve wishes to calculate κ_A or κ_B by determining the secret key of Alice or Bob from the public keys. She knows the two composed maps of Bob and one element of Alice, as follows:

$$y_{B,1}(y) = N_1^{-1} \circ x_4(y) \quad (y \in \mathcal{S}), \tag{6}$$

$$y_{B,2}(y) = x_1 \circ x_2(y) \quad (y \in \mathcal{S}), \tag{7}$$

$$y_A = N_1^{-1} \circ x_4(x_A). \tag{8}$$

Eve's Strategy 1

If x_4 is an invertible map, $y_{B,1}$ is also invertible; thus, she attempts to determine x_A from (6) and (8):

$$x_A = y_{B,1}^{-1}(y_A) = x_4^{-1} \circ N_1 \circ N_1^{-1} \circ x_4(x_A). \tag{9}$$

Subsequently, she can calculate

$$\kappa_A = x_1 \circ x_2(x_A).$$

If x_4 is not an invertible map, she obtains a set $\{x_E \in \mathcal{S} : y_A = N_1^{-1} \circ x_4(x_E)\}$ instead of an element from (9). However, in this case, she can calculate the SSK as follows:

$$x_1 \circ x_2(x_E) = x_3 \circ x_4(x_E) = x_3 \circ N_1 \circ N_1^{-1} \circ x_4(x_E) = x_3 \circ N_1(y_A) = \kappa_B. \tag{10}$$

The second equation of (10) is obtained from the compatibility condition (5), and the final equation is obtained from the definition of κ_B (see Figure 1 or Step 5 of Section 2.4).

Eve's Strategy 2

First, Eve attempts to obtain a map N_1 from (6) and then attempts to obtain a map $x_{3,E} : \mathcal{S} \to \mathcal{S}$ that satisfies

$$x_{3,E} \circ N_1 \circ N_1^{-1} \circ x_4(y) = x_1 \circ x_2(y) \tag{11}$$

for all $y \in \mathcal{S}$. Finally, she can calculate the SSK as follows:

$$x_{3,E} \circ N_1(y_A) = x_{3,E} \circ N_1 \circ N_1^{-1} \circ x_4(x_A) = x_1 \circ x_2(x_A) = \kappa_A. \tag{12}$$

Suppose that Bob constructs $y_{B,1}$ and $y_{B,2}$ to satisfy the following two requirements:

Requirement 1. It is difficult to calculate $x_4^{-1} \circ N_1 \circ N_1^{-1} \circ x_4(y) = y$ for all $y \in \mathcal{S}$ in real time, or it is difficult to determine the invertible map of $y_{B,1}$, $x_4^{-1} \circ N_1$ in real time.

Requirement 2. It is difficult to determine the map N_1 from $y_{B,1} = N_1^{-1} \circ x_4$ in real time, or it is difficult to obtain a map $x_{3,E}$ to satisfy (11) in real time.

Then, it is difficult for Eve to obtain x_A from (9) and to proceed to (12) in real time; that is, Eve cannot obtain the SSK in real time. It should be noted that the secret key x_A that Alice selects is not strongly related to Eve's breaking complexity owing to Requirement 1. This means that Bob may take substantially more responsibility for maintaining security than Alice. In this case, Alice can select her secret key space as a small one in terms of the bit size, provided that an exhaustive search for x_A is difficult in real time, and we expect that this will reduce Alice's computational complexity for y_A and κ_A.

2.2. Methods

According to this observation, the methods that can be established for the research goal can be derived as follows:

1. Introduce algorithms from SAPKA (Section 3).
2. Estimate the breaking complexity of these algorithms under the assumption that the secret key space of Alice is smaller than that of Bob and verify whether or not Alice's small key space reduces the breaking complexity of the algorithms (Sections 4.2 and 4.3).
3. For algorithms for which Alice's small key space does not reduce the breaking complexity, estimate their computational complexity to calculate the keys of both Alice and Bob (Sections 4.4 and 4.5).
4. Construct the SAPKA subclass for which the algorithms possess the relation that the complexity of Alice for her keys is smaller than that of Bob (we call this subclass the "main SAPKA subclass" until we give definition of it) and introduce the necessary conditions for the subclass by generalizing the results of 3 (Sections 5.1 and 5.2).

In Section 6, we implement some of the algorithms in Section 3 and report the experimental results. We can concretely observe what can be offered by the algorithms of the subclass constructed in Section 5.1.

At the beginning of Section 5, we add restrictions to the SAPKA framework, particularly for the public keys of Bob, before discussing our main themes. We explain these restrictions briefly below.

2.3. Restrictions on SAPKA Framework

We have already explained the high level of generality of SAPKA. However, owing to this generality, algorithms that do not ensure secure PKA are also included in this class. We present one example as follows:

Let $\mathcal{S}$ be $\mathcal{S} := \mathbb{Z}_p$ for some prime number p. Bob selects the numbers $x_B, n_B \in \mathcal{S}$ and keeps them secret. Let the maps $x_1, x_2, x_3, x_4, N_1 : \mathcal{S} \to \mathcal{S}$ be defined, for $y \in \mathcal{S}$, as:

- $x_1(y) := x_B y$
- $x_2 := id$
- $x_3(y) := x_B y$
- $x_4(y) := id$
- $N_1(y) := n_B y$.

In this case, all keys are described as in Table 2.

Table 2. Complete list of key of example algorithm of SAPKA.

	Key	Parameter
	Secret keys	maps $x_1, x_2, x_3, x_4, N_1 : \mathcal{S} \to \mathcal{S}$ and elements $x_B, n_B \in \mathcal{S}$
Bob	Public keys	$y_{B,1}(y) = N_1^{-1} \circ x_4(y) = n_B^{-1} y \quad (y \in \mathcal{S})$ $y_{B,2}(y) = x_1 \circ x_2(y) = x_B y \quad (y \in \mathcal{S})$
	SSK	$\kappa_B = x_3 \circ x_4(x_A) = x_B x_A$
	Secret key	$x_A \in \mathcal{S}$
Alice	Public key	$y_A = N_1^{-1} \circ x_4(x_A) = n_B^{-1} x_A$
	SSK	$\kappa_A = x_1 \circ x_2(x_A) = x_B x_A$

The SSK calculations are performed as follows:

$$\kappa_A = x_1 \circ x_2(x_A) = x_B x_A$$

$$\kappa_B = x_3 \circ N_1(y_A) = x_3 \circ N_1 \circ N_1^{-1} \circ x_4(x_A) = x_B x_A.$$

In this case, Bob sending $y_{B,1}$ means that n_B^{-1} must be a public key, although this element is supposed to be a secret; otherwise, Alice cannot calculate y_A. Eve can easily calculate Alice's secret key x_A with $n_B n_B^{-1} x_A = x_A$; thus, Eve can recover the SSK by calculating $y_{B,2}(x_A) = x_1 \circ x_2(x_A)$. Without limiting this type of algorithm to be included in the SAPKA class, the main subclass that we attempt to construct in Section 5 can include weak algorithms. In this case, the subclass cannot be the one that we aim to construct.

To limit algorithms such as the above example, we add restrictions on the construction of Bob's public keys using a map referred to as a "non-easily invertible map" with the following definition.

Definition 1. *The map $f_g : \mathcal{S} \to \mathcal{S}$ is called as a non-easily invertible map if the calculation $f_g^{-1} \circ f_g(y)$ for all $y \in \mathcal{S}$ is difficult to compute in the mean of the computational complexity, and the order of complexity for computing $f_g^{-1} \circ f_g(y)$ $(y \in \mathcal{S})$ is equal to or greater than $O(2^{t(n)})$, where n is the bit size of y and $t : \mathbb{N} \to \mathbb{R}^+$ is a monotonically increasing function:*

$$t(n) = an$$

for some $a \in \mathbb{R}^+ \setminus \{0\}$.

O of this definition is Landau's big-O notation (Definition 5 in Section 4.1). We limit Bob's public keys $y_{B,1}$ and $y_{B,2}$ to be constructed as

$$y_{B,1}(y) = N_1^{-1} \circ x_4(y) = f_g \circ N_1'^{-1} \circ x_4'(y)$$

$$y_{B,2}(y) = x_1 \circ x_2(y) = f_g \circ x_1' \circ x_2'(y)$$

for certain non-easily mapped $f_g : \mathcal{S} \to \mathcal{S}$ and certain maps $x_1', x_2', x_4', N_1' : \mathcal{S} \to \mathcal{S}$. By doing so, attacks from Eve cannot be performed within a polynomial time. Furthermore, with several additional restrictions in Section 5, Requirement 2 of Section 2.1 can be achieved.

We attempt 4. of the method mentioned in Section 2.2 within this restricted SAPKA class with a non-easily invertible map (the explicit definition of this restricted SAPKA class by a non-easily invertible map is also provided Section 5).

2.4. Explicit Definition of SAPKA

We present the definition of SAPKA [18]. The SAPKA algorithms have the following common ingredients:

- a multiplicative semi-group $\mathcal{S}$ with 1;
- a set $\widehat{M_\mathcal{S}}$ of easily invertible maps $: \mathcal{S} \to \mathcal{S}$, known as *noise space*; and
- a set $M_\mathcal{S}$ of maps $: \mathcal{S} \to \mathcal{S}$.

In the above, $\mathcal{S}$ is public, and $\widehat{M_\mathcal{S}}$ and $M_\mathcal{S}$ belong to Bob's secret. From the key space,

$$\mathcal{K}_B := M_\mathcal{S} \times M_\mathcal{S} \times M_\mathcal{S} \times M_\mathcal{S} \times \widehat{M_\mathcal{S}}.$$

Bob prepares the quintuple $(x_1, x_2, x_3, x_4, N_1)$ as their secret key, and x_1, x_2, x_3, x_4 must satisfy the following condition, which allows Alice and Bob to obtain the SSK.

Definition 2. *Let $\mathcal{S}$ be a multiplicative semi-group with 1. If the functions $x_1, x_2, x_3, x_4 : \mathcal{S} \to \mathcal{S}$ satisfy the following condition for all $y \in \mathcal{S}$:*

$$x_1 \circ x_2(y) = x_3 \circ x_4(y), \tag{13}$$

the maps x_1, x_2, x_3, x_4 are said to be compatible, where $\circ$ denotes a map composition.

Definition 3. *For a multiplicative semi-group $\mathcal{S}$ with 1 and the quintuple $C := (x_1, x_2, x_3, x_4, N_1)$ $\in \mathcal{K}_B$, if the maps x_1, x_2, x_3, x_4 satisfy (13), we state that C is a member of the SAPKA class. We express this relation as:*

$$C \in SAPKA.$$

The key agreement process of SAPKA is described as follows:

Step 1B Bob prepares the maps $(x_1, x_2, x_3, x_4, N_1) \in \mathcal{K}_B$, of which x_1, x_2, x_3, x_4 satisfy (13). In this case, each of x_1, x_2, x_3, x_4, N_1 is their secret key.

Step 2B Bob constructs their public keys $y_{B,1}, y_{B,2}$ as a map for each:

$$y_{B,1} := N_1^{-1} \circ x_4$$

$$y_{B,2} := x_1 \circ x_2$$

and sends $(y_{B,1}, y_{B,2})$ to Alice.

Step 1A Alice selects her secret key x_A from $\mathcal{S}$.

Step 2A Alice calculates her public key y_A as follows:

$$y_A := y_{B,1}(x_A) = N_1^{-1} \circ x_4(x_A)$$

and sends it to Bob.

Step 3A Alice calculates the SSK denoted by κ_A as follows:

$$\kappa_A := y_{B,2}(x_A) = x_1 \circ x_2(x_A).$$

Step 3B Bob calculates the SSK denoted by κ_B as follows:

$$\kappa_B := x_3 \circ N_1(y_A) = x_3 \circ N_1 \circ N_1^{-1} \circ x_4(x_A) = x_3 \circ x_4(x_A).$$

2.5. Abstract of This Study

Here, we give an outline of this study, including research goal, short abstract of each section.

Research Goal

Construction of algorithms that possess the property that Alice's complexity for SSK is smaller than that of Bob's and introduction of SAPKA subclass that includes algorithms of the same property.

Section 3: Concrete Examples of SAPKA

We show concrete examples that how Alice and Bob calculate the SSK, respectively, when the key agreement process is asymmetric. Three examples that are normal DH of SAPKA description, noise element included type DH, and matrix type DH are shown.

Section 4: Breaking Complexity of SAPKA Algorithms

In this section, we demonstrate what problems that the algorithms of Section 3 are reduced to. Estimations of breaking complexities are done under the assumption that Alice's secret key space is smaller than Bob's secret key space. Here, we can see how problems algorithms reduced to are varied according to algebraic structure and SAPKA parameters. Finally, using some of SAPKA algorithms, we discuss how small Alice's secret key space can be while maintaining security.

Section 5: Generalization

Here, we try to construct the SAPKA subclass of property that Alice's computational complexity for SSK is smaller than that of Bob. At first, we exclude algorithms of weak security from the subclass framework as we mentioned in Section 2.3. Without this procedure, the subclass can contain algorithms of above property but sufficiently weak security so that Eve can obtain the SSK within polynomial time. After this limitation is done, we define the subclass and introducing some conditions (only necessarily conditions for algorithms into the subclass) by expanding the results of Section 4 so that it can be applied to the SAPKA framework.

3. Concrete Examples of SAPKA
3.1. DH

As demonstrated in [18], the SAPKA class includes the DH algorithm. However, the process is symmetric because, in this case, $S := \mathbb{Z}_p$, where p is a prime number (thus, S forms a finite field), $g \in S$ is the public parameter, N_1^{-1} is the identity map, and Alice can calculate the public key $y_A = x_4(x_A) = g^{x_A}$, even before receiving $y_{B,1}$. The process is asymmetric in the sense that Alice and Bob use different secret information, but it is symmetric in the sense that the two perform the same operations independently.

3.2. Noised DH (NDH)

The following example is a variant of the DH that does not improve its security substantially but is useful to illustrate the concept of SAPKA algorithms simply.

The components are the same as those in Section 3.1. Bob selects their secret keys $x_B, n_B \in S$. The key agreement process of the NDH algorithm is described by the following steps.

Step 1B Bob prepares the quintuple $C_{NDH} := (x_1, x_2, x_3, x_4, N_1) \in \mathcal{K}_B$ as follows:

- $x_1(y) := (g^{x_B})^y$
- $x_2 := id$
- $x_3(y) = y^{x_B}$
- $x_4(y) = g^y$
- $N_1(y) := y^{n_B^{-1}}$,

where y is selected arbitrarily from S.

Step 2B Bob constructs their public keys $y_{B,1}, y_{B,2}$ as a map for each

$$y_{B,1}(y) := N_1^{-1} \circ x_4(y) = N_1^{-1}(g^y) = (g^y)^{n_B} = g^{n_B y}$$

$$y_{B,2}(y) := x_1 \circ x_2(y) = (g^{x_B})^y$$

and sends $(y_{B,1}, y_{B,2})$ to Alice. This is equivalent to sending $g^{n_B}, g^{x_B} \in S$.

Step 1A Alice selects her secret key x_A from S.

Step 2A Alice calculates her public key y_A as follows,

$$y_A := y_{B,1}(x_A) = N_1^{-1} \circ x_4(x_A) = g^{n_B x_A}$$

and sends it to Bob.

Step 3A Alice calculates the SSK denoted by κ_A as in the DH case:

$$\kappa_A := y_{B,2}(x_A) = x_1 \circ x_2(x_A) = (g^{x_B})^{x_A} = g^{x_B x_A}.$$

Step 3B Bob calculates the SSK denoted by κ_B as follows:

$$\begin{aligned}
\kappa_B &:= x_3 \circ N_1(y_A) \\
&= x_3 \circ N_1 \circ N_1^{-1} \circ x_4(x_A) \\
&= x_3 \circ N_1(g^{n_B x_A}) \\
&= x_3(g^{n_B^{-1} n_B x_A}) \\
&= (g^{x_A})^{x_B} \\
&= g^{x_B x_A}.
\end{aligned}$$

3.3. Schur Exponentiation-Based DH (SEDH)

Let S be the multiplicative semi-group $S := \mathcal{M}(d, \mathbb{Z}_p)$ of $d \times d$ matrices with entries in $\mathbb{Z}_p$. The algorithm introduced in this section uses *Schur exponentiation*; that is, element-wise matrix exponentiation. The symbol $c^{\circ M}$ is defined as follows:

$$\left(c^{\circ M}\right)_{ij} := \begin{cases} c^M & \text{if } M \text{ is a scalar} \\ c^{M_{ij}}_{ij} & \text{if } M \text{ is a matrix} \end{cases} \quad ; \quad i, j \in \{1, \cdots, d\}.$$

The key agreement process is as follows:

Step 1B Bob selects a matrix $x_B \in S$, an invertible matrix $N_B \in S$, and a primitive element g of $\mathbb{Z}_p$. Subsequently, he constructs the quintuple $C_{SE} := (x_1, x_2, x_3, x_4, N_1) \in \mathcal{K}_B$ for all $a, b \in \{1, \ldots, d\}$ as follows:

- $x_1(y)_{a,b} := \prod_{l \in \{1,\ldots,d\}} (g^{\circ x_B})_{a,l}^{(y)_{l,g}}$

- $x_2 := id$
- $x_3(y)_{a,b} := \prod_{l \in \{1,\ldots,d\}} (y)_{l,b}^{(x_B)_{a,l}}$
- $x_4(y) := g^{\circ y}$
- $N_1(y)_{a,b} := \prod_{l \in \{1,\ldots,d\}} (y)_{l,b}^{(N_B^{-1})_{a,l}}$,

where $y \in \mathcal{S}$. The compatibility condition holds for all $y \in \mathcal{S}$ and $a,b \in \{1,\ldots,d\}$:

$$x_1 \circ x_2(y)_{a,b} = \prod_{l \in \{1,\ldots,d\}} (g^{\circ x_B})_{a,l}^{(y)_{l,b}} = \prod_{l \in \{1,\ldots,d\}} g^{(x_B)_{a,l}(y)_{l,b}}$$

$$= g^{\sum_{l \in \{1,\ldots,d\}} (x_B)_{a,l}(y)_{l,b}} = g^{(x_B y)_{a,b}} = (g^{\circ x_B y})_{a,b}$$

$$x_3 \circ x_4(y)_{a,b} = \prod_{l \in \{1,\ldots,d\}} (g^{\circ y})_{l,b}^{(x_B)_{a,l}} = \prod_{l \in \{1,\ldots,d\}} g^{(y)_{l,b}(x_B)_{a,l}}$$

$$= g^{\sum_{l \in \{1,\ldots,d\}} (x_B)_{a,l}(y)_{l,b}} = g^{(x_B y)_{a,b}} = (g^{\circ x_B y})_{a,b}.$$

Step 2B Bob constructs their public keys $y_{B,1}, y_{B,2}$ as a map for each

$$y_{B,1}(y)_{a,b} := N_1^{-1} \circ x_4(y)_{a,b} = \prod_{l \in \{1,\ldots,d\}} (g^{\circ y})_{l,b}^{(N_B)_{a,l}}$$

$$y_{B,2}(y)_{a,b} := x_1 \circ x_2(y)_{a,b} = \prod_{l \in \{1,\ldots,d\}} (g^{\circ x_B})_{a,l}^{(y)_{l,b}}$$

for all $a,b \in \{1,\ldots,d\}$. In this case, $y_{B,1}(y)_{a,b}$ can also be expressed as follows:

$$y_{B,1}(y)_{a,b} = \prod_{l \in \{1,\ldots,d\}} (g^{\circ y})_{l,b}^{(N_B)_{a,l}} = \prod_{l \in \{1,\ldots,d\}} (g^{\circ N_B})_{a,l}^{(y)_{l,b}}.$$

Bob sends the maps $(y_{B,1}, y_{B,2})$ to Alice. This is equivalent to sending the matrices $g^{\circ N_B}, g^{\circ x_B} \in \mathcal{S}$.

Step 1A Alice selects her secret key $x_A \in \mathcal{S}$.

Step 2A Alice calculates her public key y_A for all $a,b \in \{1,\ldots,d\}$ as follows:

$$y_{A_{a,b}} := y_{B,1}(x_A)_{a,b} = N_1^{-1} \circ x_4(x_A)_{a,b} = \prod_{l \in \{1,\ldots,d\}} (g^{\circ N_B})_{a,l}^{(x_A)_{l,b}} \tag{14}$$

$$= (g^{\circ N_B x_A})_{a,b}$$

and sends it to Bob.

Step 3A Alice calculates the SSK denoted by κ_A for all $a,b \in \{1,\ldots,d\}$ as follows:

$$\kappa_{A_{a,b}} := y_{B,2}(x_A)_{a,b} := x_1 \circ x_2(x_A)_{a,b} = \prod_{l \in \{1,\ldots,d\}} (g^{\circ x_B})_{a,l}^{(x_A)_{l,b}} = (g^{\circ x_B x_A})_{a,b}. \tag{15}$$

Step 3B Bob calculates the SSK denoted by κ_B for all $a,b \in \{1,\ldots,d\}$ as follows:

$$\kappa_{B_{a,b}} := x_3 \circ N_1(y_A)_{a,b} = x_3 \circ N_1 \circ N_1^{-1} \circ x_4(x_A)_{a,b} = x_3 \circ N_1(g^{\circ N_B x_A})_{a,b}$$

$$= \prod_{l \in \{1,\ldots,d\}} (g^{\circ N_B x_A})_{l,b}^{(x_B N_B^{-1})_{a,l}} = g^{\sum_{l \in \{1,\ldots,d\}} (N_B x_A)_{l,b}(x_B N_B^{-1})_{a,l}} = g^{(x_B N_B^{-1} N_B x_A)_{a,b}}$$

$$= g^{(x_B x_A)_{a,b}} = (g^{\circ x_B x_A})_{a,b}. \tag{16}$$

This process is illustrated in Figure 2.

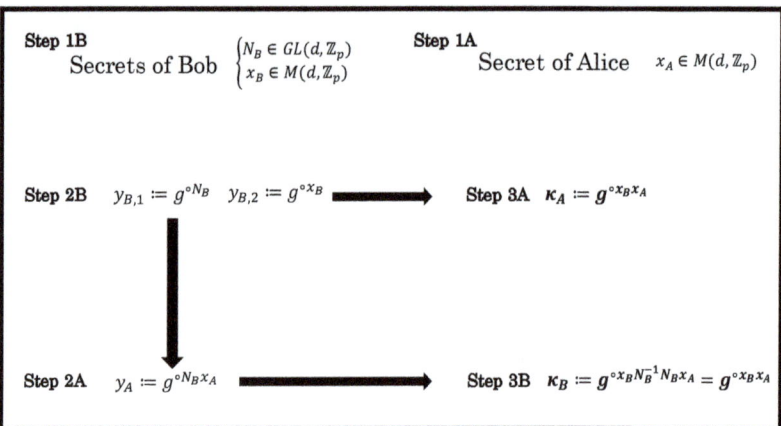

Figure 2. Key agreement process of Schur exponentiation-based DH.

Remark 1. Let $\mathcal{V}$ be a vector space of $\mathcal{V} := \mathbb{Z}_p^d$. Each of the above maps x_1, x_2, x_3, x_4, N_1 can also be considered as the map $\mathcal{V} \to \mathcal{V}$ and the condition

$$x_1 \circ x_2(y)_a = \prod_{l \in \{1,\ldots,d\}} (g^{\circ x_B})_{a,l}^{(y)_l} = (g^{\circ x_B y})_a = \prod_{l \in \{1,\ldots,d\}} (g^{\circ y})_l^{(x_B)_{a,l}} = x_3 \circ x_4(y)_a. \quad (17)$$

is satisfied for all $y \in \mathcal{V}$ and $a \in \{1,\ldots,d\}$, where x_B, N_B are the same as above. Thus, Alice can select her secret key x_A from $\mathcal{V}$ and the computational complexity for calculating y_A, κ_A, and κ_B is obviously reduced compared with the case when $x_A \in \mathcal{S} = \mathcal{M}(d, \mathbb{Z}_p)$. The precise breaking and computational complexities for the C_{SE} of both the $x_A \in \mathcal{V}$ and $x_A \in \mathcal{S}$ cases are investigated in the following section.

4. Breaking Complexity of SAPKA Algorithms

As mentioned in Section 2.2, we estimate the breaking complexity (Sections 4.2 and 4.3) and computational complexity (Section 4.4) of the algorithms in Sections 3.2 and 3.3 under the assumption in Section 4.1.

4.1. Assumptions

Let n and m be numbers in $\mathbb{N}$, and $n > m$. We define two functions $|\cdot| : \mathbb{N} \to \mathbb{N}$ and $|\cdot|_\mathcal{S} : \mathcal{S} \to \mathbb{N}$, which represent the bit lengths of the input. The difference between the two functions is the domains. The input of the first one is from $\mathbb{N}$, and the second one is from $\mathcal{S}$. The semi-groups of the above two algorithms are constructed based on a prime number p, where $|p| = n$. In this case, we can express the bit size of the element in $\mathcal{S}$ as $|y|_\mathcal{S} \leq n$ when $\mathcal{S} = \mathbb{Z}_p$ and $|y|_\mathcal{S} \leq d^2 n$ when $\mathcal{S} = \mathcal{M}(d, \mathbb{Z}_p)$. We construct subsets denoted by $\underline{\mathcal{S}}$ of $\mathcal{S}$ for both the $\mathcal{S} = \mathbb{Z}_p$ and $\mathcal{S} = \mathcal{M}(d, \mathbb{Z}_p)$ cases, respectively, as follows:

$$\underline{\mathcal{S}} := \{y \in \mathcal{S} : |y|_\mathcal{S} \leq m\} \quad \text{(when } \mathcal{S} = \mathbb{Z}_p \text{ case)}$$

$$\underline{\mathcal{S}} := \{y \in \mathcal{S} : |y|_\mathcal{S} \leq d^2 m\} \quad \text{(when } \mathcal{S} = \mathcal{M}(d, \mathbb{Z}_p) \text{ case)},$$

with the aim of determining whether the complexity of the algorithms in Sections 3.2 and 3.3 are reduced if we suppose that Alice's secret key is selected from the small bit length set $\underline{\mathcal{S}}$.

Prior to investigating the breaking complexity of the algorithms in Sections 3.2 and 3.3 under the assumption and a certain condition for n and m, we demonstrate that not only the complexity, but also the order of complexity (a definition of order is provided in Definition 5) to compute one discrete logarithm problem (DLP) differ between the cases when an exponent is selected from a larger bit length set and from a smaller bit length set. The required settings are as follows:

- a prime number p, where $|p| = n$;
- a semi-group $\mathcal{S} := \mathbb{Z}_p$;
- a subset $\underline{\mathcal{S}} := \{y \in \mathcal{S} : |y|_\mathcal{S} \leq m\} \subset \mathcal{S}$; and
- a map $x_g : \mathcal{S} \to \mathcal{S}$, $x_g(y) := g^y$, where g is a primitive element of $\mathcal{S}$.

We prove that if $n = s(m) > m$ for some monotonically increasing function $s : \mathbb{N} \to \mathbb{N}$ of not a linear and monic polynomial, the calculation of the DLP, namely

$$x_g^{-1} \circ x_g(y) = \log_g g^y = y,$$

is easier when $y \in \underline{\mathcal{S}}$ than when $y \in \mathcal{S}$ in terms of the complexity and order (Proposition 1). First, we name this input y of x_g according to the size of set y as per the following definition:

Definition 4. *Let n and m be numbers in $\mathbb{N}$ ($n > m$), and let n be the bit size of each element of $\mathbb{Z}_p$. For a given map $x_g : \mathbb{Z}_p \to \mathbb{Z}_p$, defined as*

$$x_g(y) := g^y,$$

*where g is an element in $\mathcal{S}$, if the set y belongs to $\underline{\mathcal{S}}$, we call this y an m-**bit logarithm**. If the set is $\mathcal{S}$, y is known as an n-**bit logarithm**.*

The complexity of obtaining the n-bit logarithm refers to the complexity of calculating $x_g^{-1} \circ x_g(y)$ when $y \in \mathcal{S}$ (for the m-bit logarithm, $y \in \underline{\mathcal{S}}$).

We represent the time complexity using the function $T : \mathbb{N} \to \mathbb{R}^+$, where the input is the bit size and the output is the multiplication steps. The complexity of calculating the DLP of the m-bit logarithm is given by $T(m) = 2^{\frac{m}{2}}$ [19,20]. In this case, the complexity of calculating the n-bit logarithm is $T(n) = T(s(m)) = 2^{\frac{s(m)}{2}}$. Obviously, $T(n) > T(m)$ when $n = s(m) > m$; however, we should also consider how the complexity increases as m increases. We can compare the growth rate of $T(m)$ and $T(s(m))$ using Landau's big-O notation with the following definition:

Definition 5. *For functions $T, g : \mathbb{N} \to \mathbb{R}^+$, if there exist $c, m_0 \in \mathbb{N}$ such that*

$$T(m) \leq cg(m) \tag{18}$$

is satisfied for all $m \geq m_0$, we state that $T(m)$ has an order of $g(m)$ time complexity, and we describe it as

$$T(m) \in O(g(m)). \tag{19}$$

Relation (19) indicates that $O(g)$ is a set defined by:

$$O(g) := \{T : \text{there exists } c, m_0 \in \mathbb{N} \text{ to satisfy (18) for all } m \geq m_0\}.$$

Therefore, for two complexity functions $T, T' : \mathbb{N} \to \mathbb{R}^+$, where $T(m) \leq T'(m)$ for all $m \in \mathbb{N}$ and $t(m) \in O(g(m))$, $T'(m) \in O(g'(m))$, if we wish to state that the growth rate of T' is higher than that of T (the order of T' is higher than T), we can simply describe it as follows:

$$O(g'(m)) \subset O(g(m))$$

or

$$g'(m) > g(m)$$

for all m. In this case,
$$T'(n) \notin O(g(n))$$
$$T(n) \in O(g(n)), \; T(n) \in O(g'(n))$$

hold. We can prove that the growth rate of $t(s(m)) = 2^{\frac{s(m)}{2}}$ is higher than $T(m) = 2^{\frac{m}{2}}$ as follows:

Proposition 1. *If and only if the monotonically increasing function $s : \mathbb{N} \to \mathbb{N}$ for satisfying $n = s(m) > m$ is a not linear and monic polynomial, the growth rate of $T((s(m)) = 2^{\frac{s(m)}{2}}$ is higher than $T(m) = 2^{\frac{m}{2}}$; that is,*
$$O(2^{\frac{s(m)}{2}}) \subset O(2^{\frac{m}{2}})$$

holds.

Proof. Note that $T(s(m)) \in O(2^{\frac{s(m)}{2}})$ and $T(m) \in O(2^{\frac{m}{2}})$. Suppose that s is a one-degree but non-monic polynomial defined as:
$$n = s(m) = hm + b,$$

where $h, b \in \mathbb{N}, h \neq 1$. If $O(2^{\frac{m}{2}}) = O(2^{\frac{s(m)}{2}})$ holds in this case, there must exist the constants $c, m_0 \in \mathbb{N}$ such that for all $m \geq m_0$,
$$2^{\frac{s(m)}{2}} = 2^{\frac{hm+b}{2}} \leq c 2^{\frac{m}{2}} \tag{20}$$

is satisfied. However, in this case, $2^{\frac{m(h-1)+b}{2}} \leq c$ holds, so that c cannot be a constant. Thus, in this case,
$$O(2^{\frac{s(m)}{2}}) \subset O(2^{\frac{m}{2}})$$

holds. When the case s is over a one-degree polynomial, it is obvious that there is no $c, m_0 \in \mathbb{N}$ to satisfy (20) for all $m \in m_0$.

If s is a linear and monic polynomial described as
$$n = s(m) = m + b,$$

where $b \in \mathbb{N}$, we have
$$O(2^{\frac{s(m)}{2}}) = O(2^{\frac{m+b}{2}}) = O(2^b 2^{\frac{m}{2}}) = O(2^{\frac{m}{2}}).$$

□

Therefore, if Alice and Bob attempt a key exchange using the DH algorithm, where Bob's secret key x_B is
$$x_B \in S$$

and Alice's secret key x_A is
$$x_A \in \underline{S},$$

Eve should attempt to obtain x_A, which is easier for her to obtain than x_B. The security of the algorithms in Sections 3.2 and 3.3 also depends on the difficulty of the DLP, but these algorithms may not be the same as DH owing to one of Bob's public keys $y_{B,1}$ and especially the map N_1. In Section 2.1, we discussed the requirement for $y_{B,1}$ to allow Alice to select her secret key from a small bit length set without reducing the breaking complexity. For both algorithms, Requirement 2 is satisfied if Bob selects n_B, as it is an n-bit logarithm for x_g with the algorithm of Section 3.2, and similarly, N_B is selected with the algorithm of Section 3.3. Of course, in this case, the number n must be sufficiently large to ensure that the calculation of the n-bit logarithm is not achieved in real time.

At this point, we focus on verifying whether the other algorithms in Sections 3.2 and 3.3 satisfy Requirement 1. of Section 2.1. Subsequently, we attempt to determine the polynomial s; that is, how small m can be and the computational complexity of these algorithms for each key of Alice and Bob when m is selected to be as small as possible (Section 4.4).

4.2. Eve Attempts an Attack against NDH

We assume the following:

- a prime number p, where $|p| = n$;
- a semi-group $\mathcal{S} := \mathbb{Z}_p$;
- a subset $\underline{\mathcal{S}} := \{y \in \mathcal{S} : |y|_\mathcal{S} \leq m\} \subset \mathcal{S}$;
- $x_A \in \underline{\mathcal{S}}$;
- $n_B, x_B \in \mathcal{S}$;
- a map $x_g : \mathcal{S} \to \mathcal{S}$, $x_g(y) := g^y$, where g is a primitive element of $\mathcal{S}$;
- a non-linear and monic polynomial s, where $n = s(m) > m$ for all $m \in \mathbb{N}$.

Eve's goal is to obtain x_A from the following public keys

- $y_{B,1}(y) = N_1^{-1} \circ x_4(y) = N_1^{-1}(g^y) = (g^y)^{n_B} = g^{n_B y}$
- $y_{B,2}(y) := x_1 \circ x_2(y) = (g^{x_B})^y$,

and public elements

- g^{x_B}
- g^{n_B}
- $y_A = N_1^{-1} \circ x_4(x_A) = g^{n_B x_A}$.

She must calculate

$$y_{B,1}^{-1}(y_A) = x_4^{-1} \circ N_1(y_A) = \log_g (y_A)^{n_B^{-1}} = \log_g g^{n_B x_A n_B^{-1}} = \log_g g^{x_A} = x_A. \quad (21)$$

As she does not know n_B, it appears that she must obtain the n-bit logarithm n_B by calculating $\log_g g^{n_B} = n_B$ before obtaining the m-bit logarithm x_A. However, (21) can be described as

$$y_{B,1}^{-1}(y_A) = x_4^{-1} \circ N_1(y_A) = n_B^{-1} \log_g y_A = n_B^{-1} \frac{\log_{g^{n_B}} g^{n_B x_A}}{\log_{g^{n_B}} g} = \log_{g^{n_B}} g^{n_B x_A} = x_A.$$

Thus, when we define a map $x_{g'} : \mathcal{S} \to \mathcal{S}$ such that

$$x_{g'}(y) := x_{g^{n_B}}(y) = (g^{n_B})^y,$$

$y_{B,1}^{-1} = x_{g'}^{-1}$, we can construct $x_{g'}$ using only g^{n_B}, which is public. For this attack, Eve only needs to calculate $x_{g'}^{-1}(y_A) = x_{g'}^{-1} \circ x_{g'}(x_A)$ to obtain the m-bit logarithm x_A. Therefore, it can be said that Alice selecting a small bit length set $\underline{\mathcal{S}}$ reduces the breaking complexity for Eve; that is, Requirement 1 of Section 2.1 is not satisfied in this case.

4.3. Eve Attempts an Attack against SEDH

We assume the following:

- a prime number p, where $|p| = n$;
- a semi-group $\mathcal{S} := \mathcal{M}(d, \mathbb{Z}_p)$;
- a subset $\underline{\mathcal{S}} := \{y \in \mathcal{S} : |y|_\mathcal{S} \leq d^2 m\} \subset \mathcal{S}$;
- $x_A \in \underline{\mathcal{S}}$;
- $N_B, x_B \in \mathcal{S}$;
- a map $x_g : \mathcal{S} \to \mathcal{S}$, $x_g(y) := g^y$, where g is a primitive element of $\mathcal{S}$; and
- a not linear monic polynomial s, where $n = s(m) > m$ for all $m \in \mathbb{N}$.

Eve's goal is to obtain x_A from the following public keys:

- $y_{B,1}(y)_{a,b} := N_1^{-1} \circ x_4(y)_{a,b} = \prod_{l \in \{1,\dots,d\}} (g^{\circ y})_{l,b}^{(N_B)_{a,l}}$

- $y_{B,2}(y)_{a,b} := x_1 \circ x_2(y)_{a,b} = \prod_{l \in \{1,\ldots,d\}} (g^{\circ x_B})_{a,l}^{(y)_{l,b}}$,

where $a, b \in \{1, \ldots, d\}$, and the following elements:

- $g^{\circ x_B}$
- $g^{\circ N_B}$
- $y_A = N_1^{-1} \circ x_4(x_A) = g^{\circ N_B x_A}$,

to calculate

$$y_{B,1}^{-1}(y_A)_{a,b} = x_4^{-1} \circ N_1(g^{\circ N_B x_A})_{a,b} = \log_g(g^{\circ N_B^{-1} N_B x_A})_{a,b} = \log_g(g^{x_A})_{a,b} = (x_A)_{a,b} \quad (22)$$

for all $a, b \in \{1, \ldots, d\}$. It appears that she first needs to obtain an element N_B by calculating $\log_g(g^{\circ N_B})_{a,b}$ for all $a, b \in \{1, \ldots, d\}$, the complexity of which is equivalent to the complexity for calculating d^2 n-bit logarithms for the map x_g, because she does not know N_B, so as to obtain N_B^{-1}. After obtaining N_B, she can proceed to the final two equalities of (22), and the complexity of calculating $\log_g(g^{x_A})_{a,b} = (x_A)_{a,b}$ for all $a, b \in \{1, \ldots, d\}$ is equivalent to the complexity of calculating d^2 m-bit logarithms for the map x_g. We now verify that Eve really needs to calculate at least the same amount as to obtain d^2 n-bit logarithms by calculating $\log_g(g^{\circ N_B})_{a,b}$ for all $a, b \in \{1, \ldots, d\}$. In Attack 1, we attempt to find other descriptions for map $y_{B,1}$, as in Section 4.2. In Attack 2, we attempt an attack in which Eve avoids calculating d^2 DLPs for N_B, and we compare the breaking complexity of the attack with an exhaustive search for $x_A \in \mathcal{S}$. In Attack 3, we attempt to determine the least complexity that Eve requires to obtain x_A. Hereafter, we refer to the logarithm of matrix $M \in \mathcal{M}(d, \mathbb{Z}_p)$ as follows:

$$\log_g M = (\log_g M_{a,b}) \qquad (g \in \mathbb{Z}_p)$$

$$\log_G M = (\log_{G_{a,b}} M_{a,b}) \qquad (G \in \mathcal{M}(d, \mathbb{Z}_p)).$$

Attack 1

Using the known element $g^{\circ N_B}$, Eve attempts the following calculation for all $a, b \in \{1, \ldots, d\}$:

$$\log_{g^{\circ N_B}}(y_A) = \log_{g^{\circ N_B}} g^{\circ N_B x_A}. \quad (23)$$

However, unlike in Section 4.2, the relation $\log_{g^{\circ N_B}} g^{\circ N_B x_A} = x_A$ does not hold. She obtains $X \in \mathcal{S}$ instead of $x_A \in \mathcal{S}$ from (23):

$$\log_{g^{\circ N_B}} g^{\circ N_B x_A} = \log_{g^{\circ N_B}} (g^{\circ N_B}) \circ X = X.$$

In this case, X holds for the relation

$$N_B \bullet X = N_B x_A,$$

where $\bullet$ denotes element-wise matrix multiplication. Obviously, $X = x_A$ does not always hold, so $N_1^{-1} \circ x_4(y_A) = \log_{g^{\circ N_B}}(y_A)$ also does not hold. Moreover, to obtain x_A, Eve needs to calculate $\log_g g^{\circ N_B} = N_B$.

Attack 2

We introduce an attack in which Eve does not calculate $\log_g g^{\circ N_B} = N_B$. The public key of Alice y_A is described as

$$y_A = \begin{pmatrix} \prod_{l \in \{1,\ldots,d\}} (g^{\circ N_B})_{1,l}^{(x_A)_{l,1}} & \cdots & \prod_{l \in \{1,\ldots,d\}} (g^{\circ N_B})_{1,l}^{(x_A)_{l,d}} \\ \vdots & \ddots & \vdots \\ \prod_{l \in \{1,\ldots,d\}} (g^{\circ N_B})_{d,l}^{(x_A)_{l,1}} & \cdots & \prod_{l \in \{1,\ldots,d\}} (g^{\circ N_B})_{d,l}^{(x_A)_{l,d}} \end{pmatrix}$$

$$= \begin{pmatrix} (g^{\circ N_B})_{1,1}^{(x_A)_{1,1}} (g^{\circ N_B})_{1,2}^{(x_A)_{2,1}} \cdots (g^{\circ N_B})_{1,d}^{(x_A)_{d,1}} & \cdots & (g^{\circ N_B})_{1,1}^{(x_A)_{1,d}} (g^{\circ N_B})_{1,2}^{(x_A)_{2,d}} \cdots (g^{\circ N_B})_{1,d}^{(x_A)_{d,d}} \\ \vdots & \ddots & \vdots \\ (g^{\circ N_B})_{d,1}^{(x_A)_{1,1}} (g^{\circ N_B})_{d,2}^{(x_A)_{2,1}} \cdots (g^{\circ N_B})_{d,d}^{(x_A)_{d,1}} & \cdots & (g^{\circ N_B})_{d,1}^{(x_A)_{1,d}} (g^{\circ N_B})_{d,2}^{(x_A)_{2,d}} \cdots (g^{\circ N_B})_{d,d}^{(x_A)_{d,d}} \end{pmatrix}. \quad (24)$$

Step 1 For row $b \in \{1, \ldots, d\}$ of y_A, Eve creates a matrix $W_b = (w_{a,l}^{(b)}) \in \mathcal{M}(d, \mathbb{Z}_p)$ to satisfy

$$\begin{pmatrix} (g^{\circ N_B})_{1,1}^{(x_A)_{1,b}} (g^{\circ N_B})_{1,2}^{(x_A)_{2,1}} \cdots (g^{\circ N_B})_{1,d}^{(x_A)_{d,b}} \\ \vdots \\ (g^{\circ N_B})_{d,1}^{(x_A)_{1,b}} (g^{\circ N_B})_{d,2}^{(x_A)_{2,1}} \cdots (g^{\circ N_B})_{d,d}^{(x_A)_{d,b}} \end{pmatrix} = \begin{pmatrix} w_{1,1}^{(b)} w_{1,2}^{(b)} \cdots w_{1,d}^{(b)} \\ \vdots \\ w_{d,1}^{(b)} w_{d,2}^{(b)} \cdots w_{d,d}^{(b)} \end{pmatrix}.$$

Step 2 For all $a, l \in \{1, \ldots, d\}$, Eve calculates

$$v_{a,l} := \log_{(g^{\circ N_B})_{a,l}} w_{a,l}^{(b)}. \quad (25)$$

Step 3 For all $l \in \{1, \ldots, d\}$, Eve verifies whether or not

$$v_{1,l} = v_{2,l} = \cdots = v_{d,l} \quad (26)$$

is true.

Step 4 If (26) is true for all $l \in \{1, \ldots, d\}$, Eve can recover row $b \in \{1, \ldots, d\}$ of x_A:

$$(x_A)_{1,b} = v_{1,1} = \cdots = v_{d,1}$$
$$(x_A)_{2,b} = v_{1,2} = \cdots = v_{d,2}$$
$$\vdots$$
$$(x_A)_{d,b} = v_{1,d} = \cdots = v_{d,d}.$$

As N_B is an invertible matrix, there exists exactly one $(X_{1,b}, X_{2,b}, \ldots, X_{d,b})^t \in \mathbb{Z}_p^d$ to satisfy system (24) for a given y_A, $g^{\circ N_B}$ of row $b \in \{1, \ldots, d\}$; thus, it must be $(x_{A1,b}, x_{A2,b}, \ldots, x_{Ad,b})^t$. If (26) is false in some $l \in \{1, \ldots, d\}$, Eve returns to Step 1.

In Step 2, Eve needs to calculate at least two DLPs of m-bit logarithms; thus, she requires $2 \cdot 2^{\frac{m}{2}}$ multiplications. As $\mathbb{Z}_p$ forms a group under multiplication, for any arbitrary element e in $\mathbb{Z}_p$, there exists a unique $w_{a,d}^{(b)} \in \mathbb{Z}_p$ such that:

$$ew_{a,d}^{(b)} = (y_A)_{a,b}$$

for each $a \in \{1, \ldots, d\}$. Thus, Eve has $2^{n(d-1)}$ choices for $(w_{a,1}^{(b)}, w_{a,2}^{(b)}, \ldots, w_{a,d-1}^{(b)})$ and each $a \in \{1, \ldots, d\}$, and therefore, Eve must repeat Steps 2 to 4 at most $2^{nd(d-1)}$ times. The total multiplications required for this attack is denoted as:

$$2^{nd(d-1)+\frac{m}{2}+1} = 2^{s(m)d(d-1)+\frac{m}{2}+1},$$

which is much larger than the total multiplications required for an exhaustive search on each row of $x_A \in \mathcal{S}$, namely:

$$m2^{dm}$$

Attack 3

Suppose that Eve attempts to determine x_A directly from the formula of Alice's public key y_A:

$$(y_A)_{a,b} = \prod_{l \in \{1, \ldots, d\}} (g^{\circ N_B})_{a,l}^{(x_A)_{l,b}}. \quad (27)$$

As N_B is an invertible matrix, this is equivalent to solving the following system for $X \in \underline{S}$:

$$(y_A)_{a,b} = \prod_{l \in \{1,\ldots,d\}} (g^{\circ N_B})_{a,l}^{(X)_{l,b}} \tag{28}$$

for all $a, b \in \{1, \ldots, d\}$ and given y_A, $g^{\circ N_B}$. The least complexity for finding X from (28) is expressed by the following theorem:

Theorem 1. *The complexity of solving* (28) *for* $X \in \underline{S}$ *is equal to or greater than the complexity of obtaining* N_B *from* $\log_g g^{\circ N_B}$.

Proof. Suppose that Eve can determine $X = x_A \in \underline{S}$ from (28) for a given $g^{\circ N_B}$ and y_A with a complexity denoted by the function $T : \mathbb{N} \to \mathbb{R}^+$ defined as $T(m) := p(m)$, where p is a polynomial. Under this assumption, Eve can solve system (28) even when $x_A \in S$ (thus, $X \in S$) because, in this case, the complexity is $T(s(m)) = p(s(m))$, which is also a polynomial. It should be noted that solving (28) is equivalent to solving the following system for all $a, b \in \{1, \ldots, d\}$:

$$(y_A)_{a,b}^t = \prod_{l \in \{1,\ldots,d\}} (g^{\circ N_B})^t {}_{l,b}^{(X)_{a,l}^t} \tag{29}$$

for given matrices $(g^{\circ N_B})^t$, $(y_A)^t$, where t denotes matrix transposition, because

$$(y_A)_{a,b}^t = g^{\circ(N_B x_A)^t} = g^{\circ(x_A)^t(N_B)^t} = g^{\sum_{l \in \{1,\ldots,d\}} (x_A)_{a,l}^t (N_B)_{l,b}^t}$$

$$= \prod_{l \in \{1,\ldots,d\}} (g^{\circ N_B})^t {}_{l,b}^{(x_A)_{a,l}^t}$$

for all $a, b \in \{1, \ldots, d\}$. This emphasizes that Eve can solve the system of both variable matrices by left multiplication and right multiplication on the Schur exponent of a given matrix. Therefore, once she has obtained $X \in S$ and calculates $g^{\circ X}$ within polynomial time, the following system can also be solved for an unknown $N_B \in S$:

$$(y_A)_{a,b} = \prod_{l \in \{1,\ldots,d\}} (g^{\circ X})_{l,b}^{(N_B)_{a,l}} \tag{30}$$

for all $a, b \in \{1, \ldots, d\}$ within polynomial time. As N_B is the matrix logarithm of $g^{\circ N_B}$, we can state that, if Eve can solve system (28) for X within polynomial time, she can obtain d^2 n-bit logarithms within polynomial time. □

This theorem claims that the problem of calculating $\log_g g^{\circ N_B} = N_B$ is polynomial-time-reducible to the problem of solving system (28) for $X \in \underline{S}$.

4.4. Determination of n and m

We investigate how large the bit sizes n and m must be, as well as the computational complexity of calculating the public keys and SSK for Alice and Bob according to n and m. If the attacks on SEDH are only those in Section 4.3, Eve must execute an exhaustive search for each row of x_A (as the attack is performed by each row, Eve requires d times exhaustive searches within a vector space of size 2^{dm}) or must obtain N_B, the difficulty of which is equivalent to obtaining d^2 n-bit logarithms, as in the above attacks. We assume that attacks against SEDH are only those described in Section 4.3, and we determine the sizes of n and m. Considering that the difficulty for Eve to obtain N_B depends on the difficulty of the DLP and that of obtaining x_A directly from (24) by an exhaustive search depends on the vector space of size 2^{dm}, n and m must satisfy the requirements in Table 3.

Table 3. Requirements for n and m.

	Requirements
n	Sufficiently large for one DLP
m	2^{dm} must be sufficiently large for an exhaustive search

As mentioned in Section 4, the complexity of one n-bit logarithm denoted by the function

$$T : \mathbb{N} \to \mathbb{R}^+$$

is $T(n) = T(s(m)) = 2^{\frac{s(m)}{2}}$, and the complexity of an exhaustive search for x_A of each row denoted by the function $T' : \mathbb{N} \to \mathbb{R}^+$ is $T'(m) = 2^{dm}$. (In fact, Eve must solve d^2 DLPs or d exhaustive searches for a vector space $\mathbb{Z}_{2^m}^d$, but d^2 and d are so trivial for exponential functions that we consider these complexities for only one number.) If Alice and Bob would like the breaking complexity for one DLP and that for an exhaustive search for one row of x_A to be equalized, the polynomial $s : \mathbb{N} \to \mathbb{N}$ mentioned at the beginning of Section 4.3; that is, the relation between n and m, is obtained by the following theorem:

Theorem 2. *If the time complexity for obtaining one n-bit logarithm is equal to the complexity of an exhaustive search within a vector space of size 2^{dm}, the polynomial $s : \mathbb{N} \to \mathbb{N}$ for $m \in \mathbb{N}$ is obtained by*

$$s(m) = 2dm.$$

Proof. We obtain

$$T(n) = T(s(m)) = T'(m) \iff \frac{s(m)}{2} = dm. \tag{31}$$

□

With this relation, what does the computational complexity for Alice and Bob become?

4.5. Computational Complexity of SEDH for Alice and Bob

For a given $d \in \mathbb{N}$, approximately dm multiplications are required if the bit size of the exponent is n to calculate one scalar exponentiation. Furthermore, approximately m multiplications are required if the bit size of the exponent is m. Thus, the calculation

$$g^{\circ X} \tag{32}$$

requires approximately $d^2 s(m) = 2d^3 m$ multiplications if $X \in \mathcal{S}$. If $X \in \underline{\mathcal{S}}$, it requires approximately $d^2 m$ multiplications. Moreover, the calculation

$$\prod_{l \in \{1,\ldots,d\}} (Y_{a,l})^{(X)_{l,b}} \tag{33}$$

requires $ds(m) = 2d^2 m$ multiplications if $X \in \mathcal{S}$ and dm multiplications if $X \in \underline{\mathcal{S}}$ for each $a, b \in \{1, \ldots, d\}$. Thus, according to (14) and (15), the computational complexities required by Alice for y_A and her SSK denoted by the functions $T_{y_A}, T_{\kappa_A} : \mathbb{N} \to \mathbb{R}^+$ of m are

$$T_{y_A}(m) := d^3 m \tag{34}$$

$$T_{\kappa_A}(m) := d^3 m. \tag{35}$$

From the calculations of $g^{\circ x_B}, g^{\circ N_B}$, and (16), Bob's computational complexities for $y_{B,1}, y_{B,2}$, and their SSK, denoted by the functions $T_{y_{B,1}}, T_{y_{B,2}}, T_{\kappa_B} : \mathbb{N} \to \mathbb{R}^+$ of n, become

$$T_{y_{B,1}}(n) := d^2 n \tag{36}$$

$$T_{y_{B,2}}(n) := d^2 n \tag{37}$$

$$T_{\kappa_B}(n) := d^3 n \tag{38}$$

As $n = s(m) = 2dm$, $T_{y_{B,1}}$, $T_{y_{B,2}}$, and $T_{y_{\kappa_B}}$ can be described as functions of m as follows:

$$T_{y_{B,1}}(n) = T_{y_{B,1}}(s(m)) = d^2 s(m) = 2d^3 m$$

$$T_{y_{B,2}}(n) = T_{y_{B,2}}(s(m)) = d^2 s(m) = 2d^3 m$$

$$T_{\kappa_B}(n) = T_{\kappa_B}(s(m)) = d^3 s(m) = 2d^4 m.$$

Thus, the total complexity for Alice and Bob can be compared:

$$T_{y_A}(m) + T_{\kappa_A}(m) = 2d^3 m < 4d^3 m + 2d^4 m = T_{y_{B,1}}(s(m)) + T_{y_{B,2}}(s(m)) + T_{\kappa_B}(s(m)). \tag{39}$$

When d is given, $T_{y_A}(m) + T_{\kappa_A}(m)$ and $T_{y_{B,1}}(s(m)) + T_{y_{B,2}}(s(m)) + T_{\kappa_B}(s(m))$ have the same order of complexity. $2d^4$ and $2d^3$ are considered as coefficients.

Of course, we can consider the total complexity for Alice and Bob by the functions $T'_A, T'_B : \mathbb{N} \to \mathbb{R}^+$ of d, which are defined as

$$T'_A(d) := 2md^3$$

$$T'_B(d) := 4md^3 + 2md^4$$

for a given m. In this case, $T'_A(d) \in O(d^3)$, $T'_B(d) \in O(d^4)$, and $O(d^4) \subset O(d^3)$ are obvious. Which values of m and d should be given and which should be the input of the functions can be determined the environment in which the algorithm is used, such as the computational resources and communication infrastructure.

Remark 2. As mentioned in Remark 1, Alice can select her secret key x_A not only as a matrix, but also as a vector, whereas x_B and N_B are selected as matrices. Let $\mathcal{V}$ be a vector space of $\mathcal{V} := \mathbb{Z}_p^d$, where $|y| \leq dn$ ($y \in \mathcal{V}$) and a subset $\underline{\mathcal{V}}$ of $\mathcal{V} := \{y \in \mathcal{V} : |y| \leq dm\} \subset \mathcal{V}$, and suppose that x_A is selected from $\underline{\mathcal{V}}$. In this case, the breaking strategy against this algorithm for Eve is as follows:

- solving d^2 DLPs for N_B and d DLPs for x_A, and
- an exhaustive attack for x_A within $\underline{\mathcal{V}}$, the size of which is 2^{sm}.

That is, the breaking complexity is equivalent to the case when $x_A \in \mathcal{S}$ in terms of the order of complexity. Thus, there is also no problem in selecting n and m from the relation of Theorem 2 in this case. The computational complexities for y_A, κ_A, and κ_B, which are denoted by the functions $T_{y_A, vec}, T_{\kappa_A, vec}$, and $T_{\kappa_B, vec} : \mathbb{N} \to \mathbb{R}^+$, are

$$T_{y_A, vec}(m) := d^2 m$$

$$T_{\kappa_A, vec}(m) := d^2 m$$

$$T_{\kappa_B, vec}(n) := d^2 n$$

for a given d. It can be observed that the complexities are reduced from (34), (35), and (38). Moreover, the calculation of (33) is independent for each element $a \in \{1, \ldots, d\}$, and thus, it can be computed in parallel. The calculation is also parallelized for each $l \in \{1, \ldots, d\}$. In Section 6, we report the results of the implementational experiments of SEDH in the case of $x_A \in \underline{\mathcal{V}}$. To determine how rapidly the calculation is performed, particularly for Alice, we compare the calculation speed with that of the usual DH algorithm.

5. Generalization

In this section, all investigations are conducted under the following assumptions:

- the numbers $n, m \in \mathbb{N}$ and a not linear and monic polynomial s, where $n = s(m) > m$ for all $m \in \mathbb{N}$;

- a number p, where $|p| = n$;
- a semi-group $\mathcal{S}$, where $|y|_\mathcal{S} \leq kn$ ($y \in \mathcal{S}, k \in \mathbb{N}$);
- a subset $\underline{\mathcal{S}} := \{y \in \mathcal{S} : |y|_\mathcal{S} \leq km\} \subset \mathcal{S}$; and
- $x_A \in \underline{\mathcal{S}}$.

In the above, $|\cdot|_\mathcal{S}$ is the same notation as that in Section 4.1. Furthermore, the constant k is uniquely determined by the properties of $\mathcal{S}$, such as the dimensions of the matrices and vectors. For example, $k = 1$ when $\mathcal{S} = \mathbb{Z}_p$ and $k = d^2$ when $\mathcal{S} = \mathcal{M}(d, \mathbb{Z}_p)$.

As mentioned in Section 2.3, we first construct a SAPKA subclass, in which the security of each algorithm is ensured by a non-easily invertible map (Definition 1). Thereafter, we attempt to construct the main SAPKA subclass mentioned in Section 2.2. We introduce inclusion relations between other subclasses and necessary conditions for algorithms to belong to the main subclass in Section 5.2.

We present the definition of the subclass based on the non-easily invertible map.

Definition 6. *For a multiplicative semi-group $\mathcal{S}$ and the quintuple $C := (x_1, x_2, x_3, x_4, N_1)$, where the maps x_1, x_2, x_3, x_4 satisfy (13) and N_1 is an easily invertible map of $N_1 : \mathcal{S} \to \mathcal{S}$, if maps $x'_1, x'_2, x'_3, x'_4, N'_1 : \mathcal{S} \to \mathcal{S}$ exist, where N'_1 is easily invertible and a non-easily invertible map $f_g : \mathcal{S} \to \mathcal{S}$ exists such that the following equations:*

$$x'_1 \circ x'_2(y) = x'_3 \circ x'_4(y) \tag{40}$$

$$x_1 \circ x_2(y) = f_g \circ x'_1 \circ x'_2(y) \tag{41}$$

$$N_1^{-1} \circ x_4(y) = f_g \circ N_1'^{-1} \circ x'_4(y) \tag{42}$$

$$x'_1 \circ x'_2(e) \notin \underline{\mathcal{S}} \ \lor \ N_1'^{-1} \circ x'_4(e) \notin \underline{\mathcal{S}} \tag{43}$$

are satisfied for all $y \in \mathcal{S}$, where e is an identity element of $\mathcal{S}$, the quintuple C is a member of the $SAPKA_{f_g}$ class, and we express this relation as

$$C \in SAPKA_{f_g}.$$

From (40) and (41), and the compatibility condition of (13), the equation

$$x_1 \circ x_2(y) = f_g \circ x'_1 \circ x'_2(y) = f_g \circ x'_3 \circ x'_4(y) = x_3 \circ x_4(y) \tag{44}$$

for all $y \in \mathcal{S}$ is automatically satisfied. Using (42), the equation

$$f_g \circ x'_3 \circ N'_1 \circ N_1'^{-1} \circ x'_4(y) = f_g \circ x'_3 \circ x'_4(y)$$
$$= x_3 \circ x_4(y) = x_3 \circ N_1 \circ N_1^{-1} \circ x_4(y) = x_3 \circ N_1 \circ f_g \circ N_1'^{-1} \circ x'_4(y) \tag{45}$$

holds for all $y \in \mathcal{S}$. Thus, the public keys and SSK of Alice and Bob for the algorithms in this class can be described as Table 4.

Table 4. Public keys and SSK of Alice and Bob for $SAPKA_{f_g}$ class.

	Key	Parameter
Bob	Public keys	$y_{B,1}(y) = N_1^{-1} \circ x_4(y) = f_g \circ N_1'^{-1} \circ x'_4(y)$ $(y \in \mathcal{S})$, $y_{B,2}(y) = x_1 \circ x_2(y) = f_g \circ x'_1 \circ x'_2(y)$ $(y \in \mathcal{S})$
	SSK	$\kappa_B = x_3 \circ N_1(y_A) = f_g \circ x'_3 \circ N'_1 \circ N_1'^{-1} \circ x'_4(x_A) \because (44)$
Alice	Public key	$y_A = N_1^{-1} \circ x_4(x_A) = f_g \circ N_1'^{-1} \circ x'_4(x_A)$
	SSK	$\kappa_A = x_1 \circ x_2(x_A) = f_g \circ x'_1 \circ x'_2(x_A)$

Moreover, (41) and (42) ensure that the maps $x'_1 \circ x'_2$ and $N'^{-1}_1 \circ x'_4$ and thus the element x_A, cannot be disclosed by Eve within polynomial time. If $x'_1 \circ x'_2(e) \in \mathcal{S}$ and $N'^{-1}_1 \circ x'_4(e) \in \mathcal{S}$, the complexities for calculating

$$f_g^{-1} \circ f_g \circ x'_1 \circ x'_2(e) = x'_1 \circ x'_2(e)$$

$$f_g^{-1} \circ f_g \circ N'^{-1}_1 \circ x'_4(e) = N'^{-1}_1 \circ x'_4(e)$$

are reduced compared to the case when (43) holds (see Remark 3 in Section 5.1). If $x'_1, x'_2, x'_3, x'_4, N'_1$ are linear maps, for example, Eve can obtain a linear map $x'_{3,E} \circ N'_{1,E} : \mathcal{S} \to \mathcal{S}$ to satisfy:

$$x'_{3,E} \circ N'_{1,E} \circ N'^{-1}_1 \circ x'_4(e) = x'_1 \circ x'_2(e) \tag{46}$$

by matrix multiplication, because $x'_1 \circ x'_2(e)$ is described as a matrix. Moreover, as $\mathcal{S}$ is a multiplicative semi-group, $x'_1 \circ x'_2(y)$ for any $y \in \mathcal{S}$ can be expressed as $x'_1 \circ x'_2(ye)$ and either of the equations

$$x'_1 \circ x'_2(ye) = yx'_1 \circ x'_2(e) = yx'_{3,E} \circ N'_{1,E} \circ N'^{-1}_1 \circ x'_4(e) = x'_{3,E} \circ N'_{1,E} \circ N'^{-1}_1 \circ x'_4(ye)$$

or

$$x'_1 \circ x'_2(ye) = x'_1 \circ x'_2(e)y = x'_{3,E} \circ N'_{1,E} \circ N'^{-1}_1 \circ x'_4(e)y = x'_{3,E} \circ N'_{1,E} \circ N'^{-1}_1 \circ x'_4(ye)$$

holds for all $y \in \mathcal{S}$. Thus, only if Eve knows the calculation rule of map $x_3 \circ N_1$ can she obtain a map $x_{3,E} \circ N_{1,E}$ to satisfy

$$x_{3,E} \circ N_{1,E} \circ f_g \circ N'^{-1}_1 \circ x'_4(y) = f_g \circ x'_{3,E} \circ N'_{1,E} \circ N'^{-1}_1 \circ x'_4(y) = f_g \circ x'_1 \circ x'_2(y)$$

for all $y \in \mathcal{S}$. Finally, she can calculate the SSK as follows:

$$x_{3,E} \circ N_{1,E}(y_A) = x_{3,E} \circ N_{1,E} \circ f_g \circ N'^{-1}_1 \circ x'_4(x_A) = f_g \circ x'_1 \circ x'_2(x_A) = \kappa_A.$$

Of course, if the maps $x'_1, x'_2, x'_3, x'_4, N'_1$ are not linear, this attack may be impossible, but if (43) holds and n is sufficiently large to make the calculation of $f_g^{-1} \circ f_g(y)$ ($y \notin \mathcal{S}$) difficult in real time, Eve cannot proceed to (46). Thus, Requirement 2 of Section 2.1 is achieved for all algorithms in the $SAPKA_{f_g}$ class.

Algorithms of $SAPKA_{f_g}$ Class

It can be proven that C_{NDH} in Section 3.2 and C_{SE} in Section 3.3 belong to the $SAPKA_{f_g}$ class. For C_{NDH}, if $\mathcal{S} := \mathbb{Z}_p$ and Bob defines $x'_1, x'_2, x'_3, x'_4, N'_1, f_g$ as

- $x'_1(y) := x_B y$
- $x'_2 := id$
- $x'_3(y) := x_B y$
- $x'_4(y) := id$
- $N'_1 := n_B^{-1} y$
- $f_g(y) := g^y$,

then (40) holds, and (41) and (42) are satisfied according to the descriptions of the public keys of Bob in Table 5.

Table 5. Public keys of Bob for C_{NDH}.

Public keys of Bob	$y_{B,1}(y) = N_1^{-1} \circ x_4(y) = g^{n_B y} = f_g \circ N'^{-1}_1 \circ x'_4(y)$ ($y \in \mathcal{S}$), $y_{B,2}(y) = x_1 \circ x_2(y) = g^{x_B y} = f_g \circ x'_1 \circ x'_2(y)$ ($y \in \mathcal{S}$)

For C_{SE}, if $\mathcal{S} := \mathcal{M}(d, \mathbb{Z}_p)$ and Bob defines $x'_1, x'_2, x'_3, x'_4, N'_1, f_g$ as

- $x'_1(y) := x_B y$
- $x'_2 := id$
- $x'_3(y) := x_B y$
- $x'_4(y) := id$
- $N'_1 := N_B^{-1} y$
- $f_g(y) := g^{\circ y}$,

(40) holds, and (41) and (42) are satisfied according to Table 6.

Table 6. Public keys of Bob for C_{SE}.

Public keys of Bob	$y_{B,1}(y) = N_1^{-1} \circ x_4(y) = g^{\circ N_B y} = f_g \circ N_1'^{-1} \circ x'_4(y)\ (y \in \mathcal{S})$, $y_{B,2}(y) = x_1 \circ x_2(y) = g^{\circ x_B y} = f_g \circ x'_1 \circ x'_2(y)\ (y \in \mathcal{S})$

At this point, we construct a $SAPKA_{f_g}$ subclass, the algorithms of which possess the same property as the SEDH that was investigated in Sections 4.3 and 4.4, namely biased computational complexity to Bob (see Section 5.1). We mainly focus on this class in this study.

5.1. Notations and Definitions

Prior to constructing the main subclass (Definition 9), we present several definitions and notations below. Definition 7 provides one $SAPKA_{f_g}$ subclass in which the algorithms are symmetric and include DH. In Section 5.2, we investigate how this subclass relates to the main subclass of Definition 9. Definition 8 provides the name for the input y of a given map $f'_g : \mathcal{S} \to \mathcal{S}$ according to the set of y. We use this definition to emphasize the difference in complexity for calculating $f'^{-1}_g \circ f'_g(y)$ of a given map f'_g when $y \in \underline{\mathcal{S}}$ and when $y \in \mathcal{S}$. The notations directly below Definition 8 are functions that demonstrate the complexity for calculating each key. As these notations and Definition 8 allow us to express the breaking and computational complexity for each key quantitatively, the main subclass can be defined by equalities and inequalities.

Definition 7. *For the quintuple C such that $C \in SAPKA_{f_g}$, if $N'_1 = x'_4 = id$, we state that C is a member of the symmetric $SAPKA_{f_g}$ class, and we express this relation as*

$$C \in SAPKA_{f_g, symmetry}.$$

If $C \in SAPKA_{f_g, symmetry}$, one of the public keys of Bob $y_{B,1}$ is described as

$$y_{B,1}(y) = N_1^{-1} \circ x_4(y) = f_g \circ N_1'^{-1} \circ x'_4(y) = f_g(y),$$

where $y \in \mathcal{S}$. This means that Bob sending $y_{B,1}$ is equivalent to simply informing Alice of the construction rule of y_A. Alice can calculate y_A without any secret/public information from Bob. Symmetry means that the number of public keys that both Alice and Bob construct is essentially one, and they can construct their public keys without any public/secret information from the other.

For C_{NDH} in Section 3.2, if the parameter n_B is equal to 1, C_{NDH} describes the usual DH. We denote this case of C_{NDH} as C_{DH}, and we obtain

$$C_{DH} \in SAPKA_{f_g, symmetry}.$$

Moreover, C_{SE} in the case of $N_B = E$, denoted by $C_{SE,(N_B=E)}$, is

$$C_{SE,(N_B=E)} \in SAPKA_{f_g, symmetry}.$$

Definition 8. *For a given non-easily invertible map $f'_g : \mathcal{S} \to \mathcal{S}$, if the set to which the input y belongs is $\underline{\mathcal{S}}$, we refer to the input y as a km-bit f'^{-1}_g element. If the set to which the input y belongs is $\mathcal{S}$, we refer to the input y as a kn-bit f'^{-1}_g element.*

The complexity of obtaining the kn-bit f'^{-1}_g element means the complexity of calculating $f'^{-1}_g \circ f_g(y)$ when $y \in \mathcal{S}$ (for the km-bit f'^{-1}_g element, $y \in \underline{\mathcal{S}}$).

We define the following functions of bit size that demonstrate the complexity of calculating each public key and SSK as well as the complexity for Eve to obtain a certain key:

- $T_{y_{B,1}} : \mathbb{N} \to \mathbb{R}^+$: the complexity required for Bob to construct $y_{B,1}$;
- $T_{y_{B,2}} : \mathbb{N} \to \mathbb{R}^+$: the complexity required for Bob to construct $y_{B,2}$;
- $T_{y_A} : \mathbb{N} \to \mathbb{R}^+$: the complexity required for Alice to calculate y_A;
- $T_{\kappa_A} : \mathbb{N} \to \mathbb{R}^+$: the complexity required for Alice to calculate her SSK κ_A;
- $T_{\kappa_B} : \mathbb{N} \to \mathbb{R}^+$: the complexity required for Bob to calculate their SSK κ_B;
- $T_{Eve\ x_A} : \mathbb{N} \to \mathbb{R}^+$: the least complexity required for Eve to obtain Alice's secret key x_A from public informations, where the input of $T_{Eve\ x_A}$ is m; and
- $T_{f'^{-1}_g} : \mathbb{N} \to \mathbb{R}^+$: the complexity required to compute $f'^{-1}_g \circ f_g(y)$ for map f'_g, where the input of $T_{f'^{-1}_g}$ is m if $y \in \underline{\mathcal{S}}$ and n if $y \in \mathcal{S}$.

Furthermore, for $T_{Eve\ x_A}, T_{f'^{-1}_g}$, we define the following functions to describe their order:

- $h_{Eve\ x_A} : \mathbb{N} \to \mathbb{R}^+$ to obtain the relation $T_{Eve\ x_A} \in O(h_{Eve\ x_A})$, and
- $h_{f'^{-1}_g} : \mathbb{N} \to \mathbb{R}^+$ to obtain the relation $T_{f'^{-1}_g} \in O(h_{f'^{-1}_g})$.

Example 1.

- If $T_{y_{B,1}}, T_{y_{B,2}}$, and T_{κ_B} are defined as functions of n, T_{y_A} and T_{κ_A} are defined as functions of m, and the polynomial s is concretely determined, we can compare the total computational complexity for Alice and Bob as per Section 4.5.
- The least complexity for Eve to obtain x_A in Section 3.3 is determined as

$$T_{Eve\ x_A}(m) \geq d^2 2^{\frac{s(m)}{2}}.$$

- The complexity of obtaining the $d^2 n$-bit f^{-1}_g element, where f_g is the non-easily invertible map of C_{SE} as defined in Table 6, is expressed as

$$T_{f^{-1}_g}(n) = T_{f^{-1}_g}(s(m)) = d^2 2^{\frac{s(m)}{2}}.$$

- The complexity of obtaining the $d^2 m$-bit f^{-1}_g element is determined as

$$T_{f^{-1}_g}(m) = d^2 2^{\frac{m}{2}}.$$

In general, the complexity of obtaining the kn-bit f'^{-1}_g element for a given non-easily invertible map $f'_g : \mathcal{S} \to \mathcal{S}$ is described by $T_{f'^{-1}_g}(n) = T_{f'^{-1}_g}(s(m))$, and that of the km-bit f'^{-1}_g element is $T_{f'^{-1}_g}(m)$. We can compare not only the complexity but also the order of complexity of the two using the map $h_{f'^{-1}_g}$ defined above, as per the following Remark 3.

Remark 3. *When the order of complexity for computing the km-bit f'^{-1}_g element for map f'_g is equal to $O(2^{am})$, where $a \in \mathbb{R}^+ \setminus \{0\}$:*

$$T_{f'^{-1}_g}(s(m)) \in O(h_{f'^{-1}_g}(s(m))) = O(2^{as(m)}) \subseteq O(2^{am}) = O(h_{f'^{-1}_g}(m)) \ni T_{f'^{-1}_g}(m) \quad (47)$$

holds, and the equality holds if and only if s is a linear monic polynomial for some $t \in \mathbb{N}$:

$$n = s(m) = m + t.$$

This can be proven in a similar manner to Proposition 1, and in this case, s is restricted as a non-linear and monic polynomial, so the equality never holds.

When the order of complexity for computing the km-bit $f_g'^{-1}$ element is greater than $O(2^{am})$, although the sufficient and necessary condition of s for the equality of (47) may not be the same,

$$O(h_{f_g'^{-1}}(s(m))) = O(2^{as(m)}) \subset O(2^{am}) = O(h_{f_g'^{-1}}(m)) \tag{48}$$

holds because, for all m, the ratio $\dfrac{h_{f_g'^{-1}}(s(m))}{h_{f_g'^{-1}}(m)}$ is much larger and cannot be a constant in this case. This means that when s is a not linear and monic polynomial, the problem of obtaining the kn-bit $f_g'^{-1}$ element by computing $f_g'^{-1} \circ f_g'(y)$ ($y \in \mathcal{S}$) and the problem of obtaining the km-bit $f_g'^{-1}$ element by computing $f_g'^{-1} \circ f_g'(y)$ ($y \in \underline{\mathcal{S}}$) belong to different orders of complexity classes. Furthermore, needless to say, $T_{f_g'^{-1}}(s(m)) > T_{f_g'^{-1}}(m)$ holds.

Definition 9. *For the quintuple C of $C \in SAPKA_{f_g}$, if the following relations hold:*

$$T_{Eve\ x_A}(m) \geq T_{f_g^{-1}}(s(m)) \tag{49}$$

$$T_{y_{B,1}}(s(m)) + T_{y_{B,2}}(s(m)) + T_{\kappa_B}(s(m)) > T_{y_A}(m) + T_{\kappa_A}(m), \tag{50}$$

we state that C is a member of the class that is computationally biased to Bob $SAPKA_{f_g}$, and we write this relation as

$$C \in SAPKA_{f_g, A<B}.$$

If the relation

$$T_{Eve\ x_A}(m) \leq T_{f_g^{-1}}(m) \tag{51}$$

holds, C is a member of the computationally unbiased $SAPKA_{f_g}$ class, and we write this relation as

$$C \in SAPKA_{f_g, A=B}.$$

Relation (49) indicates that the breaking complexity of an algorithm that is constructed from $C \in SAPKA_{f_g, A<B}$ is equal to or greater than the complexity that is required to obtain the kn-bit f_g^{-1} element, even if x_A is selected from $\underline{\mathcal{S}}$. Furthermore, (51) indicates that when $C \in SAPKA_{f_g, A=B}$, the breaking complexity is equal to or less than that of obtaining the km-bit f_g^{-1} element. As noted in Remark 3, not only does $T_{f_g^{-1}}(s(m)) > T_{f_g^{-1}}(m)$ hold, but the orders of complexity are also different, so there cannot exist a C' such that $C' \in SAPKA_{f_g, A<B} \cap SAPKA_{f_g, A=B}$. Thus, we obtain Lemma 1 in Section 5.2.

5.2. Inclusion Relations between Subclasses of SAPKA and Necessity Condition

Lemma 1. *For $SAPKA_{f_g, A<B} \subset SAPKA_{f_g}$ and $SAPKA_{f_g, A=B} \subset SAPKA_{f_g}$, the following relation holds:*

$$SAPKA_{f_g, A<B} \cap SAPKA_{f_g, A=B} = \emptyset.$$

According to the attack introduced in Section 4.3, we obtain the following results.

Theorem 3. *For C_{SE} in Section 3.3, if Attack 2 of Section 4.3 and the exhaustive attack for $x_A \in \underline{\mathcal{S}}$ are only attacks without calculating $\log_g g^{\circ N_B} = N_B$ or there exist other attacks but the order of complexity is equal to or greater than $O(2^{dm})$,*

$$C_{SE} \in SAPKA_{f_g, A<B}.$$

Proof. According to Attack 3 in Section 4.3 and the descriptions of $T_{Eve\ x_A}(m)$ and $T_{f_g^{-1}}(s(m))$ in Example 1, we obtain

$$T_{Eve\ x_A}(m) \geq d^2 2^{\frac{s(m)}{2}} = T_{f_g^{-1}}(s(m)).$$

Under this assumption, $n = s(m) = 2dm$ holds, and thus, we obtain

$$T_{y_{B,1}}(s(m)) + T_{y_{B,2}}(s(m)) + T_{\kappa_B}(s(m)) = 4md^3 + 2md^4 > T_{y_A}(m) + T_{\kappa_A}(m) = 2md^3$$

from Section 4.5. □

We introduce two necessary conditions for the $SAPKA_{f_g,A<B}$ class.

Theorem 4 (Necessary Condition 1). *If the quintuple $C := (x_1, x_2, x_3, x_4, N_1) \in SAPKA_{f_g,A<B}$, the relation*

$$C \notin SAPKA_{f_g,symmetry} \tag{52}$$

holds.

Proof. We prove the contrapositive. If $C \in SAPKA_{f_g,symmetry}$, Alice's public key y_A is described as:

$$y_A = N_1^{-1} \circ x_4(x_A) = f_g \circ N_1'^{-1} \circ x_4'(x_A)$$

according to (40), but N_1' and x_4' are identities; thus, y_A becomes

$$y_A = f_g(x_A).$$

Because $x_A \in \underline{S}$, x_A is a km-bit f_g^{-1} element for map f_g, and we obtain

$$T_{Eve\ x_A}(m) = T_{f_g^{-1}}(m).$$

Thus, $C \in SAPKA_{f_g,A=B}$. Moreover, Lemma 1 indicates that $C \notin SAPKA_{f_g,A<B}$. □

Corollary 1. *The quintuple C_{DH} of the usual DH is*

$$C_{DH} \in SAPKA_{f_g,A=B},$$

and $C_{SE,(N_B=E)}$ defined in Section 5.1 is

$$C_{SE,(N_B=E)} \in SAPKA_{f_g,A=B}.$$

Theorem 5 (Necessary Condition 2). *If the quintuple $C := (x_1, x_2, x_3, x_4, N_1) \in SAPKA_{f_g,A<B}$, then there does't exist a map $f_g' : S \to S$ such that to satisfy*

$$f_g \circ N_1'^{-1} \circ x_4'(y) = f_g'(y) \wedge T_{f_g'^{-1}}(m) \leq T_{f_g^{-1}}(m) \tag{53}$$

for all $y \in S$, $m \in \mathbb{N}$.

Proof. We prove the contrapositive. We assume that for $C \in SAPKA_{f_g,A<B}$, there exists a map $f_g' : S \to S$ to satisfy

$$f_g \circ N_1'^{-1} \circ x_4'(y) = f_g'(y) \tag{54}$$

$$T_{f_g'^{-1}} \leq T_{f_g^{-1}} \tag{55}$$

for all $y \in S$. According to (54), y_A can be described as

$$y_A = f_g \circ N_1'^{-1} \circ x_4'(x_A) = f_g'(x_A).$$

This indicates that x_A is a km-bit $f_g'^{-1}$ element for a map f_g'. In combination with (55), the relation

$$T_{Eve\, x_A}(m) = T_{f_g'^{-1}}(m) \leq T_{f_g^{-1}}(m)$$

holds. Thus, $C \in SAPKA_{f_g, A=B}$; that is, $C \notin SAPKA_{f_g, A<B}$ is proven. □

Corollary 2.

$$C_{NDH} \in SAPKA_{f_g, A=B}$$

Proof. The attack of Section 4.2 indicates that

$$y_A = f_g \circ N_1'^{-1} \circ x_4'(x_A) = g^{n_B x_A} = (g^{n_B})^{x_A} = f_g'(x_A),$$

where $f_g'(y) := f_{f_g(N_1'^{-1} \circ x_4'(1))}(y) = f_{g^{n_B}}(y)$, and (53) is satisfied. □

Remark 4. *Theorem 1 of Section 4.3 implies that there does not exist a map f_g' to satisfy $f_g'(y) = f_g \circ N_1'^{-1} \circ x_4'(y)$ with $T_{f_g'^{-1}} \leq T_{f_g^{-1}}$ as for algorithm of C_{SE}.*

According to the proof of Theorem 4 and Corollary 2, we obtain the following inclusion relation:

Corollary 3.

$$SAPKA_{f_g, symmetry} \subset SAPKA_{f_g, A=B}.$$

Proof. From the proof of Theorem 4,

$$C \in SAPKA_{f_g, symmetry} \Rightarrow C \in SAPKA_{f_g, A=B}$$

is obtained. Thus, we obtain

$$SAPKA_{f_g, symmetry} \subseteq SAPKA_{f_g, A=B}.$$

Moreover, C_{NDH} is related to $C_{NDH} \notin SAPKA_{f_g, symmetry}$, but $C_{NDH} \in SAPKA_{f_g, A=B}$ according to Corollary 2. Thus, $SAPKA_{f_g, symmetry} \subset SAPKA_{f_g, A=B}$ is obtained. □

Figure 3 presents the inclusion relations between the subclasses of SAPKA.

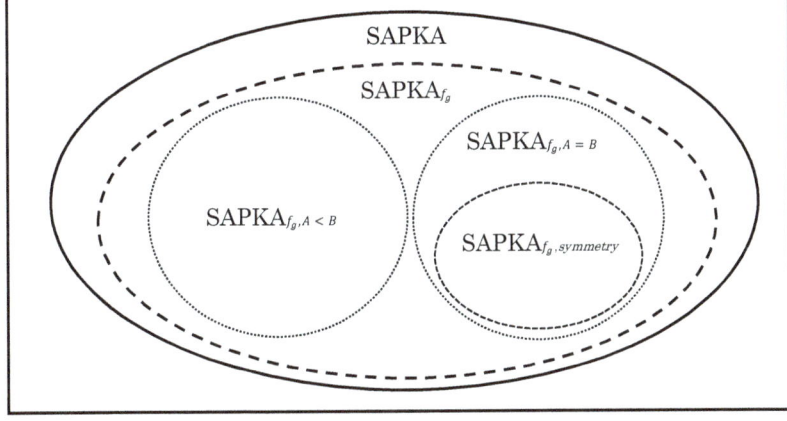

Figure 3. Inclusion relations between subclasses of SAPKA.

Remark 5. *The proposition*

$$SAPKA_{f_g,A<B} \cup SAPKA_{f_g,A=B} = SAPKA_{f_g}.$$

as well as the necessary conditions (52) *and* (53) *as sufficient conditions for the* $SAPKA_{f_g,A<B}$ *class have not yet been confirmed. These will be investigated in our subsequent paper.*

6. Implementational Experiments

In this section, we report the simple experimental results. All experiments were performed in the following environment:

- OS: Windows 10 Pro
- Processor: 2.90 GHz Intel Core i7-10700
- RAM: 8 GB
- Language: JAVA

Tables 7–9 indicate the computational complexity for Alice (y_A, κ_A) and Bob ($y_{B,1}$, $y_{B,2}$, κ_B) according to d and m while n was fixed as 1024, 3072, and 5120 bits. As can be observed from Tables 7–9, the calculation time for Bob increased as d increased. This is because the calculation complexities of (32) and (33) were dependent only on d, whereas n was fixed. For Alice's computation, there were cases in which the calculation time decreased as d increased (Tables 8 and 9). This can be explained by (32) and (33) for each $a \in \{1,\ldots,d\}$, being independent, and an increase in d resulted in a decrease in m. Therefore, these values can be computed in parallel with a smaller m when d increases. When $d > 4$, the calculation time will increase despite the decrease in the size of m owing to a shortage in processors for efficient parallel computing.

Moreover, the difference in the order of complexity for Alice and Bob can be observed from Figure 4, where m was fixed as 128 bits and d increased from 2 to 16.

Table 7. Calculation time (ms) for Alice and Bob when $n = 1024$.

d	m	Bob (ms)	Alice (ms)
2	256	7.69307551	1.89631633
4	128	16.7975061	3.39046122
6	85	25.310163	5.37513889

Table 8. Calculation times (ms) for Alice and Bob when $n = 3072$.

d	m	Bob (ms)	Alice (ms)
2	768	131.792466	17.612742
4	384	272.971624	16.480546
6	256	472.95086	28.988962

Table 9. Calculation times (ms) for Alice and Bob when $n = 5120$.

d	m	Bob (ms)	Alice (ms)
2	1280	528.017186	65.886198
4	640	1164.85195	57.475974
6	426	1942.06485	108.263714

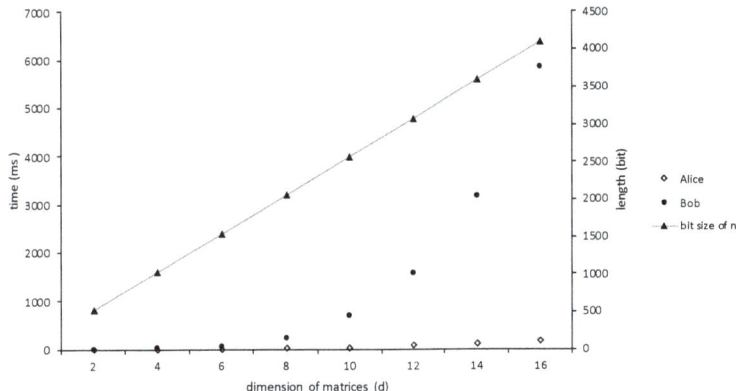

Figure 4. Comparison of calculation times (ms) for Alice and Bob according to d (m = 128).

We verified how rapidly the calculation of Alice could be achieved by comparing the speed with that of the usual DH. Table 10 and Figure 5 compare the calculation times of SEDH and DH for y_A and κ_A for the same security level (n) from 1024 to 7168 bits. The calculation speeds in the following table and figure are the averages of 50 calculations.

Table 10. Comparison of calculation times (ms) for Alice with DH.

Security (n)	(m)	(d)	DH (ms)	SEDH (ms)
1024	256	2	2.727202	1.896316327
2048	512	2	15.514816	6.451664815
3072	384	4	51.776748	16.480546
4096	1024	2	111.332034	32.10356
5120	640	4	202.806142	57.475974
6144	768	4	335.208752	94.12185
7168	896	4	535.273036	144.93926

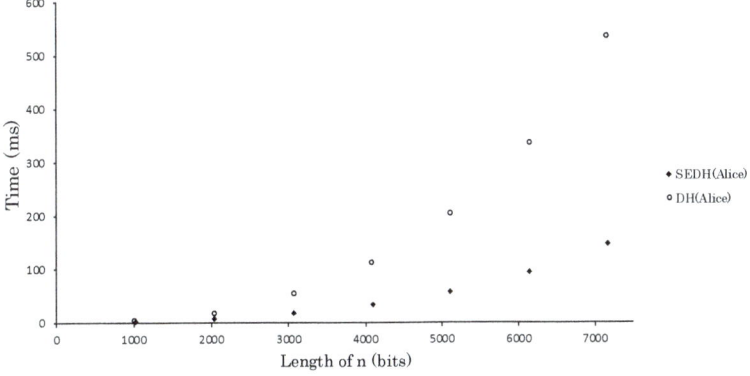

Figure 5. Comparison of calculation times (ms) for Alice with DH.

7. Conclusions

In this study, we have estimated the breaking complexity and computational complexity of two SAPKA class algorithms. One of these potentially possesses the property whereby the security is less dependent on the secret key space of Alice than that of Bob. This property allows Alice to calculate her SSK with less complexity than Bob in the algorithm.

Moreover, we generalized these algorithms and constructed SAPKA subclasses according to the security/efficiency properties. We constructed several subclass algorithms in which the above property holds. We expect that algorithms in this class will aid in constructing secure communication infrastructures, even with a small capability of devices for one side (Alice's computational asymmetry). The necessary conditions for this class and inclusion relations with other SAPKA subclasses were also investigated.

In the following Future Work section, we provide several unclear points, which include new problems obtained from this study (indicated by •) and problems not discussed in this paper but that must be investigated for practical use of presented algorithms and frameworks (indicated by ○).

Future Works

- Is it true that $SAPKA_{f_g, A<B} \cup SAPKA_{f_g, A=B} = SAPKA_{f_g}$? (Remark 5)
- Is it true that if the relations (52) and (53) hold for $C \in SAPKA_{f_g}$, C is a member of the $SAPKA_{f_g, A<B}$ class? That is, are (52) and (53) also sufficient conditions for the $SAPKA_{f_g, A<B}$ class? As mentioned in Section 2.2, this paper does not provide how to construct secure PKA algorithms resilient to any types of theoretical attacks. However, if sufficient conditions for the $SAPKA_{f_g, A<B}$ class are found, algorithms that cannot be broken at least within polynomial time can be easily constructed because those are protected thanks to non-easily mappable f_g (page 23). Of course, conditions might differ depending on required security level, but considering the sufficient conditions surely helps us provide how to construct secure PKA algorithms.
- Does S need to be selected as a multiplicative semi-group/group? For example, the algorithms in Section 3.3 are effective even in vector space, which forms an additive group (Remark 1). We should investigate how we can construct PKA algorithms on algebraic structures other than multiplicative semi-groups and then compare the breaking and computational complexities with the original one.
- We should attempt to describe algorithms such as LWE-based or NTRU-type algorithms with the SAPKA parameters and verify whether these algorithms belong to the $SAPKA_{f_g, A<B}$ class. Otherwise, would it be possible to modify these to belong to the $SAPKA_{f_g, A<B}$ class?
- It is well known that key agreement algorithms are vulnerable against man-in-the-middle (MITM) attack. Obviously, also algorithms in SAPKA are also not exceptional without any measures. Moreover, the number of public keys especially for Bob is larger than symmetric type algorithms, so they way to manage public keys for SAPKA algorithms must be discussed more. Adaptability with techniques such as digital signature [21,22] and public key certifications [23] must be well discussed.
- As mentioned in [1], the SSKs that Alice and Bob calculate should be ephemeral, and each key must have enough randomness. To practically use algorithms introduced in this paper, if public keys and SSK are sufficiently random to protect Eve from deducing SSK must be investigated.

Author Contributions: Conceptualization, K.J., S.I. and M.R.; Formal analysis, K.J. and S.I.; Implementation, K.J.; Writing—original draft, K.J.; Writing—review and editing, K.J., S.I. and M.R. All authors have read and agreed to the published version of the manuscript.

Funding: This research received no external funding.

Institutional Review Board Statement: Not applicable.

Informed Consent Statement: Not applicable.

Conflicts of Interest: The authors declare no conflict of interest.

Abbreviations

The following abbreviations are used in this manuscript:

PKA   public key agreement
SAPKA   strongly asymmetric public key agreement
SSK   secret shared key
DH   Diffie-Hellman
NDH   Noised Diffie-Hellman
SEDH   Schur Exponentiation-Based Diffie-Hellman
DLP   discrete logarithm problem
MITM   man-in-the-middle

References

1. Shannon, C.E. Communication Theory of Secrecy Systems. *Bell Syst. Tech. J.* **1949**, *28*, 656–715. [CrossRef]
2. Diffie, W.; Hellman, M. New directions in cryptography. *IEEE Trans. Inf. Theory* **1976**, *22*, 644–654. [CrossRef]
3. Rivest, R.L.; Shamir, A.; Adleman, L. A method for obtaining digital signatures and public-key cryptosystems. *Commun. ACM* **1978**, *21*, 120–126. [CrossRef]
4. Adrian, D.; Bhargavan, K.; Durumeric, Z.; Gaudry, P.; Green, M.; Halderman, J.A.; Heninger, N.; Springall, D.; Thomé, E.; Valenta, L.; et al. Imperfect Forward Secrecy: How Diffie-Hellman Fails in Practice. In Proceedings of the 22nd ACM SIGSAC Conference on Computer and Communications Security, Denver, CO, USA, 12–16 October 2015; pp. 5–17.
5. Shor, P.W. Polynomial-Time Algorithms for Prime Factorization and Discrete Logarithms on a Quantum Computer. *SIAM J. Comput.* **1997**, *26*, 1484–1509. [CrossRef]
6. Bernstein, D.J.; Chuengsatiansup, C.; Lange, T.; van Vredendaal, C. NTRU Prime: Reducing Attack Surface at Low Cost. In *Selected Areas in Cryptography—SAC 2017*; Lecture Notes in Computer Science; Adams, C., Camenisch, J., Eds.; Springer: Cham, Switzerland, 2017; Volume 10719. [CrossRef]
7. Hülsing, A.; Rijneveld, J.; Schanck, J.; Schwabe, P. High-Speed Key Encapsulation from NTRU. In *Cryptographic Hardware and Embedded Systems—CHES 2017*; Lecture Notes in Computer Science; Fischer, W., Homma, N., Eds.; Springer: Cham, Switzerland, 2017; Volume 10529. [CrossRef]
8. Hoffstein, J.; Pipher, J.; Silverman, J.H. NTRU: A ring-based public key cryptosystem. In *Algorithmic Number Theory—ANTS 1998*; Lecture Notes in Computer Science; Buhler, J.P., Ed.; Springer: Berlin/Heidelberg, Germany, 1998; Volume 1423. [CrossRef]
9. Bos, J.; Ducas, L.; Kiltz, E.; Lepoint, T.; Lyubashevsky, V.; Schanck, J.M.; Schwabe, P.; Seiler, G.; Stehle, D. CRYSTALS—Kyber: A CCA-Secure module-lattice-based KEM. In Proceedings of the 2018 IEEE European Symposium on Security and Privacy (EuroS and P), London, UK, 24–26 April 2018; pp. 353–367. [CrossRef]
10. Alkim, E.; Ducas, L.; Poppelmann, T.; Schwabe, P. Post-quantum key exchange—A new hope. In Proceedings of the 25th USENIX Security Symposium (USENIX Security 16), Austin, TX, USA, 10–12 August 2016; pp. 327–343.
11. Post-Quantum Cryptography Competition Round 2 Submissions. Available online: https://csrc.nist.gov/projects/post-quantum-cryptography/round-2-submissions (accessed on 16 March 2021).
12. PQC Standardization Process: Third Round Candidate Announcement. Available online: https://csrc.nist.gov/News/2020/pqc-third-round-candidate-announcement (accessed on 16 March 2021).
13. Micciancio, D.; Regev, O. Worst-case to average-case reductions based on Gaussian measures. *J. Comput.* **2007**, *37*, 267–302. [CrossRef]
14. Regev, O. On lattices, learning with errors, random linear codes, and cryptography. *J. ACM* **2009**, *56*, 1–40. [CrossRef]
15. Laine, K.; Lauter, K. Key Recovery for LWE in Polynomial Time. IACR Cryptology ePrint Archive 2015, no. 176. Available online: https://eprint.iacr.org/2015/176.pdf (accessed on 14 June 2021).
16. Peikert, C. Public-key cryptosystems from the worst-case shortest vector problem. In Proceedings of the forty-first annual ACM symposium on Theory of computing, Bethesda, MD, USA, 31 May–2 June 2009; pp. 333–342.
17. Coppersmith, D.; Shamir, A. Lattice Attacks on NTRU. In *Advances in Cryptology—EUROCRYPT '97*; Lecture Notes in Computer Science; Fumy, W., Ed.; Springer: Berlin, Germany, 1997; Volume 1233. [CrossRef]
18. Accardi, L.; Iriyama, S.; Regoli, M.; Ohya, M. *Strongly Asymmetric Public Key Agreement Algorithms*; Technical Report ISEC2011-20; IEICE: Tokyo, Japan, 2011; pp. 115–121.
19. Pollard, J. Monte Carlo Methods for Index Computation ($mod p$). *Math. Comput.* **1978**, *32*, 918–924.
20. Pohlig, S.; Hellman, M. An improved algorithm for computing logarithms over $GF(p)$ and its cryptographic significance (Corresp.). *IEEE Trans. Inf. Theory* **1978**, *24*, 106–110. [CrossRef]
21. Lamport, L. *Constructing Digital Signatures from a One-Way Function*; Technical Report CSL-98; SRI International: Menlo Park, CA, USA, 1979; Volume 238.

22. Merkle, R.C. *A Certified Digital Signature, Conference on the Theory and Application of Cryptology*; Springer: New York, NY, USA, 1989.
23. Cooper, D.; Santesson, S.; Farrell, S.; Boeyen, S.; Housley, R.; Polk, W. Internet X. 509 Public Key Infrastructure Certificate and Certificate Revocation List (CRL) Profile, RFC 5280. 2008; pp. 1–151. Available online: https://datatracker.ietf.org/doc/html/rfc5280 (accessed on 14 June 2021).

Article

The Design and FPGA-Based Implementation of a Stream Cipher Based on a Secure Chaotic Generator

Fethi Dridi [1,2], Safwan El Assad [2,*], Wajih El Hadj Youssef [1], Mohsen Machhout [1] and René Lozi [3]

1. Electronics and Microelectronics Laboratory (EmE), Faculty of Sciences of Monastir, University of Monastir, 5019 Monastir, Tunisia; fethi.dridi@fsm.u-monastir.tn (F.D.); wajih.hajyoussef@enim.u-monastir.tn (W.E.H.Y.); mohsen.machhout@fsm.rnu.tn (M.M.)
2. IETR (UMR 6164) Laboratory, CNRS, University of Nantes, F-44000 Nantes, France
3. J. A. Dieudonné (UMR 7351) Laboratory, CNRS, University of Cote d'Azur, 06103 Nice, France; rene.lozi@univ-cotedazur.fr
* Correspondence: safwan.elassad@univ-nantes.fr

Citation: Dridi, F.; El Assad, S.; El Hadj Youssef, W.; Machhout, M.; Lozi, R. The Design and FPGA-Based Implementation of a Stream Cipher Based on a Secure Chaotic Generator. *Appl. Sci.* **2021**, *11*, 625. https://doi.org/10.3390/app1102 0625

Received: 30 November 2020
Accepted: 5 January 2021
Published: 11 January 2021

Publisher's Note: MDPI stays neutral with regard to jurisdictional clai-ms in published maps and institutio-nal affiliations.

Copyright: © 2021 by the authors. Licensee MDPI, Basel, Switzerland. This article is an open access article distributed under the terms and conditions of the Creative Commons Attribution (CC BY) license (https:// creativecommons.org/licenses/by/ 4.0/).

Abstract: In this study, with an FPGA-board using VHDL, we designed a secure chaos-based stream cipher (SCbSC), and we evaluated its hardware implementation performance in terms of computational complexity and its security. The fundamental element of the system is the proposed secure pseudo-chaotic number generator (SPCNG). The architecture of the proposed SPCNG includes three first-order recursive filters, each containing a discrete chaotic map and a mixing technique using an internal pseudo-random number (PRN). The three discrete chaotic maps, namely, the 3D Chebyshev map (3D Ch), the 1D logistic map (L), and the 1D skew-tent map (S), are weakly coupled by a predefined coupling matrix M. The mixing technique combined with the weak coupling technique of the three chaotic maps allows preserving the system against side-channel attacks (SCAs). The proposed system was implemented on a Xilinx XC7Z020 PYNQ-Z2 FPGA platform. Logic resources, throughput, and cryptanalytic and statistical tests showed a good tradeoff between efficiency and security. Thus, the proposed SCbSC can be used as a secure stream cipher.

Keywords: chaos-based stream cipher; SPCNG; 3D chebyshev; logistic; skew-tent; FPGA; performance

1. Introduction

The protection of information against unauthorized eavesdropping and exchanges is essential, in particular for military, medical, and industrial applications. Nowadays, cryptographic attacks are more and more numerous and sophisticated; consequently, new effective and fast techniques of information protection have appeared or are under development. In this context, recent works have focused on designing new chaos-based algorithms, which provide reliable security while minimizing the cost of hardware and computing time. Chaos theory was first discovered in the computer system by Edward Lorenz in 1963 [1]. A chaotic system, although deterministic and not truly random, has unpredictable behavior, due to its high sensitivity to initial conditions and control parameters which constitute the secret key. It can generate an aperiodic analog signal whenever its phase space is continuous (i.e., with an infinity of values). However, when its phase state is discrete (with a finite set of values), its orbits must be periodic, even with a very long period.

In the field of chaos-based digital communication systems, the chaotic signal has been one of the main concerns in recent decades and is widely used to secure communication. In chaos-based cryptography, discrete chaotic maps are used in most chaotic systems (encryption, steganography, watermark, hash functions) to generate pseudo-random chaotic sequences with robust cryptographic properties [2–10]. In a stream cipher, the pseudo-random number generator (PRNG) is the most important component since all the security of the system depends on it. For this, a new category of pseudo-chaotic number generator

(PCNG) has been recently built to secure stream-data [11–14]. These PCNGs use combined chaotic maps because single chaotic maps are not secure for use in stream ciphers.

In 2017, M. Abu Taha et al. [15] designed a novel stream cipher based on an efficient chaotic generator; the results obtained from the cryptographic analysis and of common statistical tests indicate the robustness of the proposed stream cipher. In 2018, Ons et al. [16] developed two new stream ciphers based on pseudo-chaotic number generators (PCNGs) that integrate discrete chaotic maps and use the weak coupling and switching technique introduced by Lozi [17]. Indeed, the obtained results show that the proposed stream ciphers can be used in practical applications, including secure network communication.

In 2019, Ding et al. [18] proposed a new lightweight stream cipher system based on chaos—a chaotic system—and two nonlinear feedback shift registers (NFSRs) are used. The results show that the stream cipher has good cryptographic characteristics. In 2020, Abdelfatah et al. [19] proposed several efficient multimedia encryption techniques based on four combined chaotic maps (Arnold Map, Lorenz Map, Chebyshev Map, and logistic Map) using serial or parallel connections. With the rapid growth of Internet of Things (IoT) technology that connects devices with low power consumption and low computing resources, the hardware implementation of chaotic and non-chaotic ciphers is more suitable than a software implementation. Note that few chaotic encryption systems are realized in the hardware [20–22].

In this study, we designed an efficient chaos-based stream cipher (SCbSC) using a proposed secure PCNG. Then, we addressed the hardware implementation and evaluated the performance in terms of resilience against cryptanalytic attacks and in terms of hardware metrics (areas, throughput, and efficiency). The proposed system uses three weakly coupled chaotic maps (3D Chebyshev, logistic, and skew-tent) and integrates a masking technique in the recursive cells to resist side-channel attacks (SCAs). Its implementation on a Xilinx XC7Z020 PYNQ-Z2 FPGA hardware platform achieves a throughput of 1.1 Gbps at an operating frequency of 37.25 MHZ.

The main contributions of the proposed chaotic system are: First all, the introduction of some countermeasures to fix side channel attacks (SCAs) which is done using the masking technique on the recursive cells, and to fix division and conquer attacks on the initial vector (IV), using a weakly coupling matrix. Second, its hardware implementation on a Xilinx XC7Z020 PYNQ-Z2 FPGA platform and evaluation of its performance in terms of computational complexity and security.

The remainder of this paper is organized as follows. The next Section 2 presents the architecture of the proposed secure chaos-based stream cipher. Section 3 presents the hardware implementation on the Xilinx XC7Z020 PYNQ-Z2 FPGA platform of the proposed secure pseudo-chaotic number generator (SPCNG) and analyzes its performance. Section 4, investigates the performance of the proposed SCbSC in terms of hardware metrics and cryptanalytic analysis. Finally, Section 5 summarizes the whole paper.

2. The Proposed SCbSC-Based Architecture

The block diagram of a stream encryption/decryption system is presented in Figure 1. As we can see, the stream encryption/decryption algorithm comes down to an XOR operation between the plaintext and the keystream for encryption; the ciphertext and the keystream for decryption. The security of such a system depends entirely on the keystream delivered by the keystream generator. If the keystream is perfectly random and the period tends to infinity, then the encryption/decryption system becomes unconditionally secure (called a one-time pad). The keystream generator takes as input a secret key and an initial value (IV) used to overcome known plain text attacks. The IV is changed with each new session and must be used only once. Thus, the sequences generated in the different sessions with the same secret key are different. Recall that stream ciphers are used to encrypt data (bits or samples) continuously, such as network communications or selective video encryption. In the following, we will describe in detail the proposed SPCNG as a secure keystream generator.

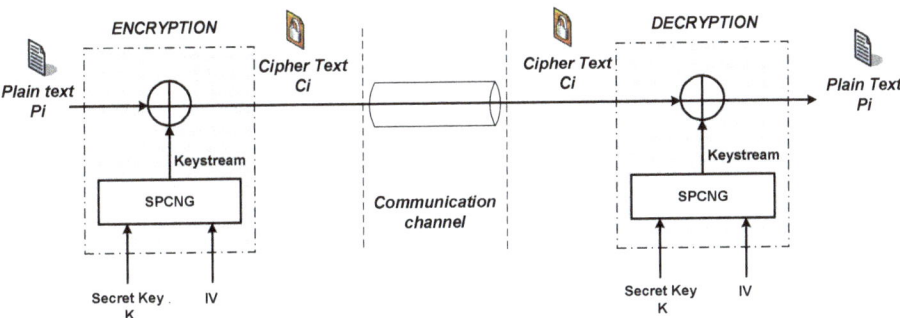

Figure 1. Block diagram of a stream encryption/decryption system.

Description of the Architecture of the Proposed SPCNG

The architecture of the proposed SPCNG is on the one hand, partly based on one of our previous PCNG [16,17], and on the other hand, it takes into account the vulnerabilities detected by SCAs [23,24] in one of our other PCNGs [15]. This new architecture makes it possible to resist SCAs. The proposed system comprises three one-delay recursive cells, shown in blue, containing weakly coupled chaotic maps, namely: the logistic map (L), the skew-tent (S), and the 3D Chebyshev map (3D Ch) in parallel with a linear feedback shift register (LFSR), and a mixing technique on each recursive cell, depicted in red, as shown in Figure 2.

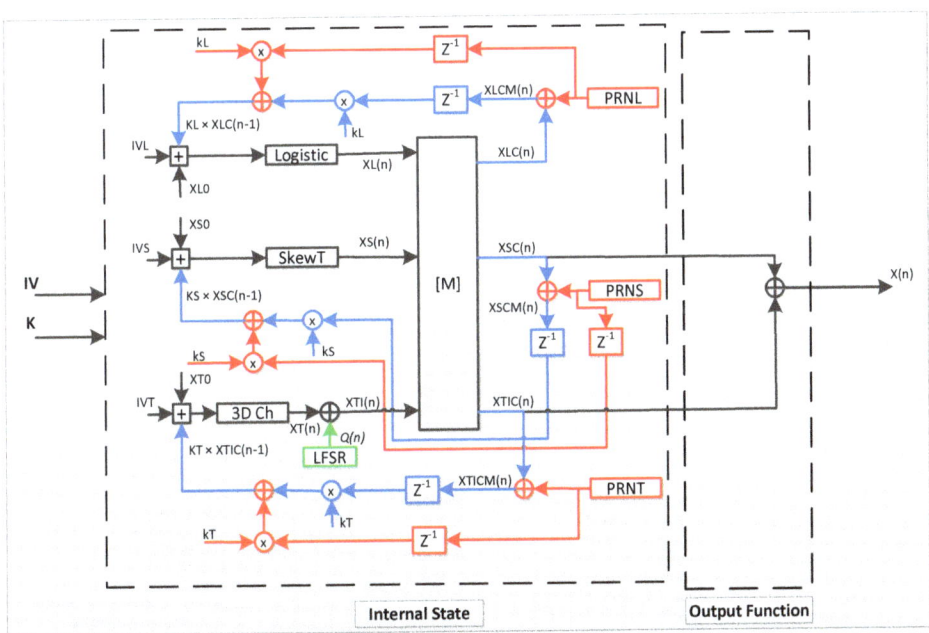

Figure 2. Architecture of the proposed SPCNG.

The M-matrix weak coupling technique creates an interdependence between the three chaotic maps that avoids an attacker using the divide and conquer approach on the first

128-bit IV. Indeed, for each new sample calculation, an attacker must take into account the three chaotic maps together. Besides, the use of the logistic map and especially the 3D Chebyshev map (which we have discretized) adds robustness to the system against algebraic attacks. Finally, the three recursive one-delay cells are protected against SCAs by using a mixing technique based on three internal pseudo-random numbers: PRNL, PRNS, and PRNT respectively, shown in red.

The proposed SPCNG takes as input an initial vector (IV) and a secret key (K). The IV of the system provides the initial vectors of the three chaotic maps, IVL, IVS, and IVT; the initial condition $XS0$ of the skew-tent map; and the initial seeds $X0\_L$, $X0\_S$, and $X0\_T$ (of 128 bits each) of the three pseudo-random numbers PRNL, PRNS, and PRNT. The output of each PRN is of size N = 32 bits. The secret key K provides the initial conditions and parameters of the SPCNG listed in Table 1.

Table 1. Composition of the secret key K.

Symbol	Definition
XL0 and XT0	The initial conditions of the chaotic maps: logistic and 3D Chebyshev respectively, ranging from 1 to $2^N - 1$.
XLC1, XSC1, and XTIC1	The initial conditions of the delayed values in recursive cells: logistic, skew-tent, and 3D Chebyshev respectively, in the range $[1, 2^N - 1]$.
Q0	The initial value Q0 of the Linear Feedback Shift Register (LFSR) defined by: $Q(n) = x^{32} + x^{22} + x^2 + x + 1.$
KL, KS, and KT	The coefficients of the recursive cells: logistic, skew-tent, and 3D Chebyshev respectively, ranging from 1 to $2^N - 1$.
P_s	The control parameter of the skew-tent map, in the range $[1, 2^N - 1]$.
ϵ_{ij}	The parameters of the coupling matrix M, in the interval $[1, 2^k]$ with $k \leq 5$.
T_r	The transient phase of 10 bits.

Note that $XLC1$, $XSC1$, and $XTIC1$ mean $XLC(-1)$, $XSC(-1)$, and $XTIC(-1)$.

The models of the discrete logistic, skew-tent, and 3D Chebyshev maps are respectively given by:

- The discrete logistic map [25]:

$$XL(n) = \begin{cases} \left\lfloor \frac{XL(n-1)[2^N - XL(n-1)]}{2^{N-2}} \right\rfloor & if\ XL(n-1) \neq [3 \times 2^{N-2}, 2^N] \\ 2^N - 1 & otherwise \end{cases} \quad (1)$$

This is the discretized equation of the standard logistic map:

$$x(n) = \mu x(n-1)(1 - x(n-1)) \quad (2)$$

with here $\mu = 4$ and $x(n) \in [0, 1]$.

- The discrete skew-tent map [26]:

$$XS(n) = \begin{cases} \left\lfloor \frac{2^N \times XS(n-1)}{P_s} \right\rfloor & if\ 0 < XS(n-1) < P_s \\ \left\lfloor 2^N \times \frac{[2^N - XS(n-1)]}{2^N - P_s} \right\rfloor & if\ P_s < XS(n-1) < 2^N \\ 2^N - 1 & otherwise \end{cases} \quad (3)$$

This is the discretized equation of the standard skew-tent map:

$$x(n) = \begin{cases} \frac{x(n-1)}{p} & if\ 0 \leq x(n-1) \leq p \\ \frac{1-x(n-1)}{1-p} & if\ p \leq x(n-1) \leq 1 \end{cases} \quad (4)$$

with $x(n) \in [0,1]$; $0 < p < 1$.

- The discrete 3D Chebyshev map [27]:

$$XT(n) = \left\lfloor 2^{(-2N+2)} \times \begin{pmatrix} 4 \times \left(XT - 2^{(N-1)}\right)^3 - 3 \\ \times 2^{(2N-2)} \times \left(XT - 2^{(N-1)}\right) \end{pmatrix} + 2^{(N-1)} \right\rfloor \quad (5)$$

This is the discretized equation of the standrd 3D Chebyshev map:

$$x(n) = 4[x(n-1)]^3 - 3x(n-1). \quad (6)$$

with $x(n) \in [-1,1]$.

$\lfloor Z \rfloor$ (floor function) is the greatest integer less than or equal to Z and $X(n)$ takes integer values $\in [1, 2^N - 1]$ and $N = 32$ is the precision used.

In Figure 3a,b, we show the mapping and attractor of the 3D Chebyshev map respectively.

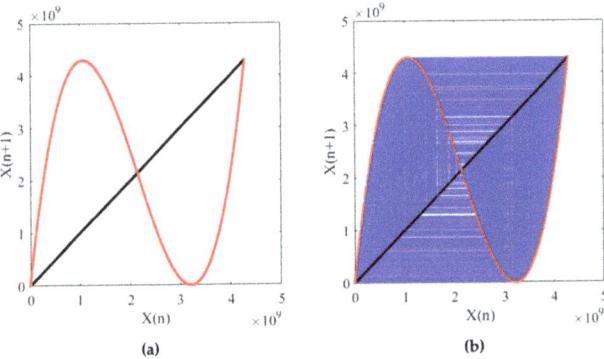

Figure 3. (a) Mapping of the 3D Chebyshev map (3D Ch); (b) its attractor.

Further, in Figure 4a,b, we give the histogram of a sequence produced by the 3D Chebyshev map alone and the histogram of a sequence generated by the 3D Chebyshev map in parallel with an LFSR, respectively. As we can see, the histogram of Figure 4b becomes uniform (confirmed by the chi-square test) compared to that of Figure 4a. The histograms of the skew-tent and logistic maps are known, and an example of their shape is given in [28].

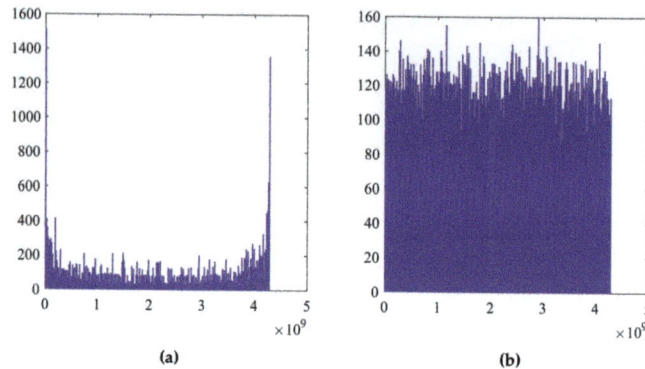

Figure 4. (a) Histogram of the 3D Ch; (b) histogram of the 3D Ch in parallel with a linear feedback shift register (LFSR).

The first sample is calculated by:

$$XL(1) = Logistic\{mod\left(XL(0), 2^N\right)\} \tag{7}$$

$$XS(1) = SkewT\{mod\left(XS(0), 2^N\right), P_s\} \tag{8}$$

$$XT(1) = 3D\,Ch\{mod\left(XT(0), 2^N\right)\} \tag{9}$$

where $XL(0)$, $XS(0)$ and $XT(0)$ are the initial values (inputs) of the three chaotic maps defined as follows:

$$\begin{array}{l} XL(0) = (IVL + XL0 + KL \times XLC1) \\ XS(0) = (IVS + XS0 + KS \times XSC1) \\ XT(0) = (IVT + XT0 + KT \times XTIC1) \end{array} \tag{10}$$

Afterward, for $n \geq 2$ and $n \leq N_s$, we calculate the samples by the following relations:

$$XL(n) = Logistic\{mod\left(KL \times XLC(n-1), 2^N\right)\} \tag{11}$$

$$XS(n) = SkewT\{mod\left(KS \times XSC(n-1), 2^N\right), P_s\} \tag{12}$$

$$XT(n) = 3D\,Ch\{mod\left(KT \times XTIC(n-1), 2^N\right)\} \tag{13}$$

where N_s is the number of the desired samples, and $XLC(n-1)$, $XSC(n-1)$, and $XTIC(n-1)$ are the unmasked inputs of the three chaotic maps.

The coupling system is defined by the following relation:

$$\begin{bmatrix} XLC(n) \\ XSC(n) \\ XTIC(n) \end{bmatrix} = M \times \begin{bmatrix} XL(n) \\ XS(n) \\ XTI(n) \end{bmatrix}, \tag{14}$$

where:

$$M = \begin{bmatrix} M_{11} & \varepsilon_{12} & \varepsilon_{13} \\ \varepsilon_{21} & M_{22} & \varepsilon_{23} \\ \varepsilon_{31} & \varepsilon_{32} & M_{33} \end{bmatrix}, \tag{15}$$

with $M_{11} = (2^N - \varepsilon_{12} - \varepsilon_{13})$, $M_{22} = (2^N - \varepsilon_{21} - \varepsilon_{23})$, and $M_{33} = (2^N - \varepsilon_{31} - \varepsilon_{32})$.

$XL(n)$, $XS(n)$, and $XT(n)$ are the outputs of the chaotic maps: logistic, skew-tent, and 3D Chebyshev respectively, and

$$XTI(n) = XT(n) \oplus Q(n) \tag{16}$$

where $Q(n)$ is the output of the LFSR.

The masking operations aim to randomize the intermediate results, and they are carried out by adding a random value to the outputs of the weak coupling samples $XLC(n)$, $XSC(n)$, and $XTIC(n)$.

$$XLCM(n) = XLC(n) \oplus PRNL(n)$$
$$XSCM(n) = XSC(n) \oplus PRNS(n) \quad (17)$$
$$XTICM(n) = XTIC(n) \oplus PRNT(n)$$

where $XLCM(n)$, $XSCM(n)$, and $XTICM(n)$ represent the masked outputs of the recursive cells: logistic, skew-tent, and 3D Chebyshev, respectively, and $PRNL(n)$, $PRNS(n)$, and $PRNT(n)$ are random integer values generated by the Xorshift pseudo-random number generator of random integer values, in the range $[1, 2^N - 1]$. To get the same output X(n) for the same secret key and the same IV, the masking operations are reversed at the inputs of the chaotic maps.

$$XLC(n-1) \times KL = XLCM(n-1) \times KL \oplus PRNL(n-1) \times KL$$
$$XSC(n-1) \times KS = XSCM(n-1) \times KS \oplus PRNS(n-1) \times KS \quad (18)$$
$$XTIC(n-1) \times KT = XTICM(n-1) \times KT \oplus PRNT(n-1) \times KT$$

Note that PRNs are based on Xoshiro's RNG, which was developed by David Blackman and Sebastiano Vigna [29] in 2019, which serves as a parameter module for PRNs. The Xoshiro construction itself is based on the Xorshift concept invented by George Marsaglia [30]. Therefore, the masking operation is an effective countermeasure to protect the implementation against power analysis-based side-channel attacks (SCAs) [31,32]. Note that the VHDL implementation of these PRNs produce 32 bits at each clock cycle.

Algorithm 1 summarizes the full operation of the proposed SPCNG.

Algorithm 1: Generation of the pseudo-chaotic sequence X(n).

Result: Keystream X(n)
initialization;
$XL(0) = (IVL + XL0 + KL \times XLC1)$;
$XS(0) = (IVS + XS0 + KS \times XSC1)$;
$XT(0) = (IVT + XT0 + KT \times XTIC1)$;
Calculation of the first sample;
$XL(1) = Logistic\{mod(XL(0), 2^N)\}$;
$XS(1) = SkewT\{[mod(XS(0), 2^N), P_s]\}$;
$XT(1) = 3D\,Ch\{mod(XT(0), 2^N)\}$;
Samples generation;
$M_{11} = (2^N - \epsilon_{12} - \epsilon_{13})$;
$M_{22} = (2^N - \epsilon_{21} - \epsilon_{23})$;
$M_{33} = (2^N - \epsilon_{31} - \epsilon_{32})$;
while $n \geq 2$ and $n < Ns$ **do**
  Internal State;
  Unmasking operations;
  $XLC(n-1) \times KL = XLCM(n-1) \times KL \oplus PRNL(n-1) \times KL$;
  $XSC(n-1) \times KS = XSCM(n-1) \times KS \oplus PRNS(n-1) \times KS$;
  $XTIC(n-1) \times KT = XTICM(n-1) \times KT \oplus PRNT(n-1) \times KT$;
  $XL(n) = Logistic\{mod(KL \times XLC(n-1), 2^N)\}$;
  $XS(n) = SkewT\{[mod(KS \times XSC(n-1), 2^N), P_s]\}$;
  $XT(n) = 3D\,Ch\{mod(KT \times XTIC(n-1), 2^N)\}$;
  $XTI(n) = XT(n) \oplus Q(n)$;
  $XLC(n) = (M_{11} \times XL) + (\epsilon_{12} \times XS)) + (\epsilon_{13} \times XTI)$;
  $XSC(n) = (\epsilon_{21} \times XL) + (M_{22} \times XS) + (\epsilon_{23} \times XTI)$;
  $XTIC(n) = (\epsilon_{31} \times XL) + (\epsilon_{32} \times XS) + (M_{33} \times XTI)$;
  SPCNG's output;
  $X(n) = XSC(n) \oplus XTIC(n)$;
end

3. Hardware Implementation of the Proposed SCbSC and Evaluation of Its Performance

The implementation of the secure chaos-based stream cipher was realized on the PYNQ Z-2 FPGA prototyping board from Xilinx. For implementation, the SCbSC's code was written in VHDL with 32-bit fixed-point data formats, then synthesized, and implemented using the Xilinx Vivado design suite (V.2017.2). Vivado design tools essentially made it possible to carry out the various steps from design to implementation on the target FPGA board. It allows, among other things, description, synthesis, simulation, and implementation of a design, then programming it on a chip from one of the different families of Xilinx FPGAs. In Figure 5, we summarize the different steps of the design flow that were performed under Vivado for the performance evaluation of the proposed SPCNG.

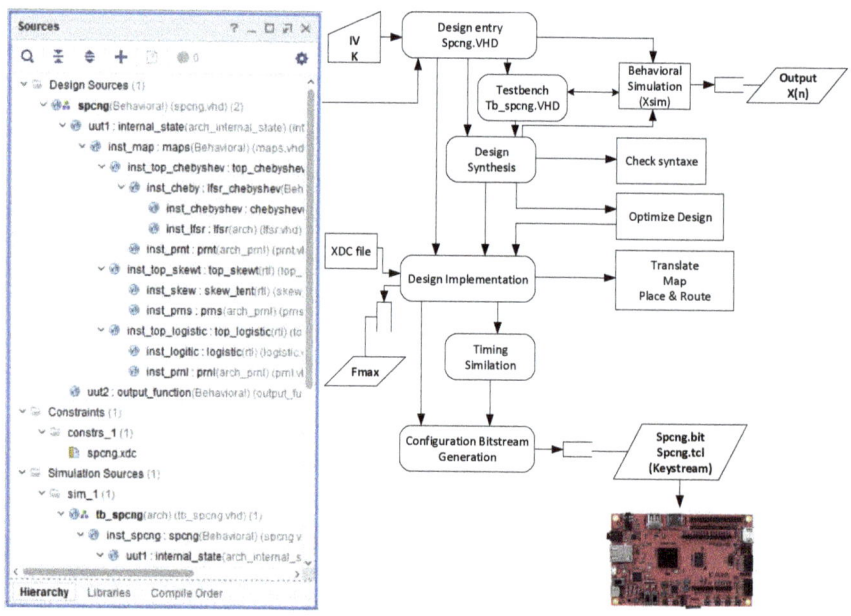

Figure 5. FPGA conception flow (under Vivado) of the proposed SPCNG.

First, we describe the proposed SPCNG using a hierarchical description containing several modules described in VHDL. Second, the synthesis step checks the VHDL description of the SPCNG, converts it into a gate-level representation, and creates a netlist. Third, we perform a behavioral simulation of the SPCNG to check its validity and make sure that the results obtained X(n) are consistent with those gotten by MATLAB. The simulation was invoked directly by the Xsim simulator integrated into the Vivado tools and the results obtained are displayed in a chronogram (see Figure 6 (Behavioral simulation)). At this step we can assess the statistical performance of the SPCNG. Fourth, the design implementation performs: First, Translate merges the netlists resulting from the design synthesis and the specified constraints file (Xilinx Design Constraint XDC file); then Map fits the design with the available resources of the target FPGA. After that, the Place and Route process places the components and routes them, respecting the constraints specified during the translation, to obtain a configuration file. At this step, we get the maximum frequency and hardware resources summarized in the implementation reports. After the design implementation, we performed the post-implementation timing simulation to get the true timing delay information of the SPCNG as shown in the chronogram of Figure 6 (Post-implementation timing simulation).

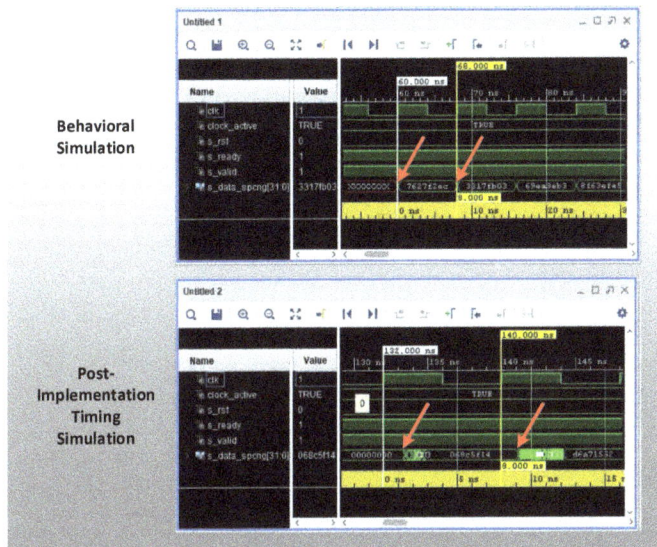

Figure 6. Behavioral simulation and Post-implementation timing simulation.

Finally, we generated a programming file (BIT) to program the Xilinx device PYNQ-Z2 FPGA.

3.1. Hardware Cost of the Proposed Secure PCNG

In this section, we analyze the performance of the proposed SPCNG implementation in terms of resources used (area, DSP), speed (maximum frequency—Max. Freq., throughput), and efficiency. Four SPNG versions were realized to choose the best among them in terms of hardware resources, throughput, and statistical resilience (NIST test) for use in the SCbSC system (see Table 2). Furthermore, we give the efficiency (in terms of throughput/slices) of all versions. The efficiency parameter gives us an overall idea of the hardware metrics performance of the implementation.

$$Max.Freq. = \frac{1}{T - WNS}[Mhz]. \qquad (19)$$

where $T = 8$ ns is the target clock period ($F = 1/T = 125$ Mhz) and WNS is the worst negative slack of the clock signal in the intra-clock paths section.

$$Throughput = N \times Max.Freq.[Mbps]. \qquad (20)$$

$$Efficiency = \frac{Throughput}{Slices}[Mbps/Slices]. \qquad (21)$$

The proposed SPCNG versions were implemented on a Xilinx XC7Z020 PYNQ-Z2 FPGA hardware platform.

Table 2. Comparison of the proposed SPCNG design versions on ZYNQ PYNQ Z2 FPGA.

			Versions			
			Chaotic Multiplexing		XOR Operation	
			Without LFSR	With LFSR	Without LFSR	With LFSR
Resources used	Area	LUTs	3744/7.04%	3763/7.07%	3586/6.74%	3599/6.77%
		FFS	1066/1%	1130/1.06%	1064/1%	1128/1.06%
		Slices *	1079/8.11%	1087/8.17%	1031/7.75%	1029/7.74%
	DSPs		25/11.36%	25/11.36%	22/10%	22/10%
Speed	WNS [ns]		−18.968	−19.062	−19.632	−18.018
	Max. Freq. [Mhz]		37.08	36.95	36.18	38.43
	Throughput [Mbps]		1186.59	1182.46	1158.07	1229.91
Efficiency [Mbps/Slices]			1.09	1.08	1.12	1.19
NIST			Successful	Successful	Successful	Successful

* Note: Each slice contains four LUTs with 6 inputs and eight FFs.

The four SPCNG versions have the same general structure but are completely different in their output function and slightly different in their internal state. The differences between the versions of columns 1 and 2 on the one hand, and the versions of columns 3 and 4 on the other hand, are in the output function used, as shown in Table 2. Indeed, versions 1 and 2 use a chaotic multiplexing technique as output function, where the sequence X(n) is controlled by a chaotic sample $X_{th}(n)$ and a threshold T_{th} is defined as follows:

$$X(n) = \begin{cases} XSC(n) & if\ 0 < X_{th} < T_{th} \\ XTIC(n) & otherwise \end{cases} \quad (22)$$

with $X_{th}(n) = XLC(n) \oplus XSC(n)$ and $T_{th} = 0.8 \times 2^N$.

Version 2, compared to version 1, contains a LFSR in parallel with the 3D Chebyshev map. Version 4 is the one shown in Figure 2, and version 3 is the same as version 4, but without the LFSR. Moreover, all SPCNG versions successfully passed the 15 NIST tests. However, versions without LFSR did not pass certain sub-tests. For the chaotic multiplexing technique, we found only one failed sub-test out of 148 non-overlapping template sub-tests, and for the XOR operation, we found three failed sub-tests out of 148 non-overlapping template sub-tests. Therefore, based on all results in Table 2, we chose version 4, which is the best (in terms of resources used, throughput, and efficiency) compared to other versions, to be used in the SCbSC system.

3.2. SPCNG Resilience against Statistical Attacks

To quantify the cryptographic properties of the pseudo-chaotic sequences generated by the proposed SPCNG, a series of tests must be applied. Each test measures a particular characteristic, such as the correlation between generated sequences or their uniformity, and the overall results of these tests give an idea of the degree of randomness of the sequences produced. The pseudo-chaotic behavior of the generated sequences is closely linked to the statistical characteristics of these sequences. The National Institute of Standards and Technology (NIST) tests [33] serve, among other things, as a reference to quantify and compare the statistical properties of binary pseudo-chaotic sequences.

Note that the Lyapunov exponents of the three chaotic maps used are positive; however, it is not obvious to compute the Lyapunov exponents of the new stream cipher we propose here. Nevertheless, its chaotic nature is due mainly to the weak coupling of the three chaotic maps. The weak coupling mechanism of chaotic maps has been thoroughly studied [17]; it leads generally to high quality pseudo-random generators. The chaotic nature of it is highlighted by the histogram and figures of the uniform and uncorrelated distribution of its iterates (Figures 7 and 8).

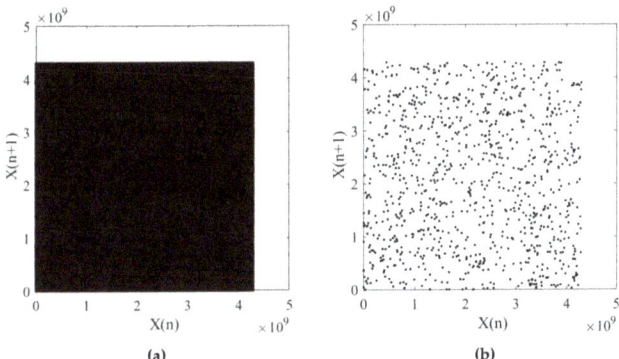

Figure 7. (**a**) Mapping of a sequence X(n) of 3,125,000 samples, generated by the proposed SPCNG and the mapping of 1000 samples taken randomly from X(n) in (**b**).

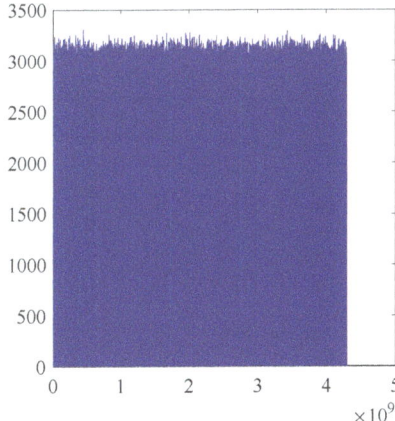

Figure 8. Histogram.

3.2.1. Phase Space Test

We draw in Figure 7a the phase space or mapping of a sequence X(n) generated by the proposed SPCNG formed by 3,125,000 samples out of the 3,125,100 samples generated to deviate from the transitional regime $T_r = 100$, and in Figure 7b, we show the mapping of 1000 samples taken randomly from $X(n)$.

Already, from Figure 7b, the region looks like a totally disordered region, indicating the lack of correlation between adjacent sample values.

3.2.2. Histogram and Chi-Square Tests

An important key property of a secure pseudo-chaotic number generator is that the sequences generated should have a uniform distribution. The histogram of a sequence X (n) produced is given in Figure 8, the uniformity of which is observed visually.

The visual uniformity result should be confirmed by the chi-square test formulated as follows:

$$\chi^2_{ex} = \sum_{i=0}^{N_c-1} \frac{(O_i - E_i)^2}{E_i} \quad (23)$$

where:

- $N_c = 1000$: number of classes.
- O_i: number of calculated samples in the ith class E_i.
- $E_i = N_s/N_c$: expected number of samples of a uniform distribution.
- Ns: the number of samples produced—here, Ns = 3,125,000

After that step, we obtain: $\chi^2_{ex} = 909.46 < \chi^2_{th}(N_c - 1; \alpha) = 1073.64$ (for $N_c = 1000$ and $\alpha = 0.05$). The experimental value of the chi-square test is less than the theoretical one, asserting the histogram's uniformity. This test was performed on 100 different sequences using 100 different secret keys, and all sequences were uniform.

3.2.3. NIST Test

Another important key property of a secure pseudo-chaotic number generator is that the sequences generated should pass the statistical NIST test, which is a package of 188 tests and sub-tests used to evaluate the randomness of long binary sequences. NIST test was applied to 100 pseudo-chaotic sequences of size 10^8 bits, generated from the initial conditions and the parameters of the chaotic system. For each test, a set of 100 p-values was calculated to indicate the result of the test. A p-value larger than $\alpha = 0.01$ (the level of significance of the test) indicates that the sequence would be random and a p-value less than 0.01 means that the sequence is nonrandom. The proportion of 100 sequences passing a test is equal to the number of p-values $\geq \alpha$ divided by 100. The results obtained, given in Table 3, indicate that the sequences generated passed all 15 statistical tests.

Table 3. *P*-values and proportion results of NIST test.

Test	*p*-Value	Proportion %
Frequency test	0.616	100
Block-frequency test	0.182	97
Cumulative-sums test (2)	0.825	99.5
Runs test	0.956	100
Longest-run test	0.868	100
Rank test	0.182	99
FFT test	0.868	99
Nonperiodic-templates (148)	0.507	98.912
Overlapping-templates	0.956	99
Universal	0.575	98
Approximate Entropie	0.658	99
Random-excursions (8)	0.511	99.432
Random-excursion-variant(18)	0.376	99.832
Serial test (2)	0.290	98
Linear-complexity	0.834	100

This means that the proposed SPCNG produces indistinguishable sequences of integer random sequences.

4. Performance Analysis of the Proposed SCbSC

In this section, we first give the hardware metrics obtained by the proposed SCbSC system and compare them with those of some published systems. Then, and we assess its security against a known cryptanalytic analysis.

4.1. SCbSC Hardware Metrics

The hardware metrics of the SCbSC system are shown in Table 4, and as expected, they are similar to those of SPCNG.

Table 4. Hardware metrics of the proposed SCbSC.

Resources used	Area	LUTs	3631/6.83%
		FFS	1225/1.15%
		Slices	1081/8.13%
	DSPs		22/10%
Speed	WNS [ns]		−18.845
	Max. Freq. [Mhz]		37.25
	Throughput [Mbps]		1192.02
Efficiency [Mbps/Slices]			1.1

The comparison of the hardware metrics of several chaotic and non-chaotic systems (from eSTREAM project phase-2 focus hardware profile) is summarized in Table 5. This comparison is difficult to interpret due to the differences in characteristics of the FPGAs tested—particularly for the clock rate parameter. However, considering the clock rate of the FPGA board and the efficiency achieved, we can make this comparison. Thus, the SCbSc system presents competitive hardware metrics compared to those obtained from most other chaotic and non-chaotic systems, except the Trivium cipher. However, since 2007, different types of attacks have been applied to eSTREAM ciphers, thereby revealing some weaknesses, in particular on Trivium cipher [34,35]. Indeed, in Trivium AND gates are the only nonlinear elements to prevent attacks that exploit, among other things, the linearity of linear feedback shift registers.

Table 5. Hardware metrics usage comparison of several chaotic and non-chaotic systems.

Cipher	Device	Frequency [Mhz]		Slices	Throughput [Mbps]	Efficiency [Mbps/slices]
		Clock Frequency	Max. Freq.			
SCbSC	Pynq Z2	125	37.25	1081	1119.02	1.1
LWCB SC [20]	Zynq7000	-	18.5	2363 LUTs	565	-
Lorenz's chaotic System [21]	Virtex-II	50	15.598	1926	124	0.06
Chaos-ring [22]	Virtex-6	125	464.688	1050	464.688	0.44
Trivium [36]	Spartan 3	50	190	388	12,160	31.34
Grain-128 [37]	Virtex- II	50	181	48	181	3.77
Mickey-128 [37]	Virtex- II	50	200	190	200	1.05

4.2. Cryptanalytic Analysis

In order to assess the security of the proposed SCbSC system against the most common attacks, we performed the following the key space analysis and assessed its sensitivity; then we used statistical analysis.

4.2.1. Key Size and Sensitivity Analysis

For a secure image encryption system, the key space should be large enough to resist a brute-force attack [38]. The secret key is produced here by Xorshif generator [30] and its size is given by:

$$|K| = |XL0| + |XT0| + |Q0| + |XLC1| + |XSC1| + |XTIC1| + |P_s| \\ + |KL| + |KS| + |KT| + |T_r| + (6 \times |\varepsilon_{ij}|) = 360 \text{ bits} \quad (24)$$

where $|XL0| = |XT0| = |Q0| = |XLC1| = |XSC1| = |XTIC1| = |P_s| = |KL| = |KS| = |KT| = 32$ bits, $|T_r| = 10$ bits, and $|\varepsilon_{ij}| = 5$ bits.

Thus, the key space contains 2^{360} different combinations of the secret key, which is large enough to make brute force attack impracticable.

A robust cryptosystem should also be sensitive to the secret key; that is, changing a one bit in the secret key must produce a completely different encrypted image. This sensitivity is conventionally measured by two parameters which are the NPCR (number of pixel change rate) and the UACI (unified average changing intensity) [39]. Besides, instead of those two parameters which operate on the bytes, we use the Hamming distance H_D which operates on the bits (in our opinion H_D is more precise than NPCR and UACI parameters). The expressions of these parameters are given below, with C_1 and C_2 being the two ciphered images of the same plain image P.

$$NPCR = \frac{1}{M \times N} \sum_{i,j} D(i,j) \times 100\% \tag{25}$$

$$D(i,j) = \begin{cases} 1 & \text{if } C_1(i,j) \neq C_2(i,j) \\ 0 & \text{if } C_1(i,j) = C_2(i,j) \end{cases} \tag{26}$$

where M and N are the width and height of C_1 and C_2. The NPCR measures the percentage of different pixel numbers between two ciphered images.

$$UACI = \frac{1}{M \times N \times 255} \sum_{i,j} |C_1(i,j) - C_2(i,j)| \times 100\% \tag{27}$$

which measures the average intensity of differences between the two images.

$$H_D(C_1, C_2) = \frac{1}{Nb} \sum_{i=1}^{Nb} (C_1(i) \oplus C_2(i)) \tag{28}$$

with Nb being the number of bits in an encrypted image.

For a random image, the expected values of NPCR, UACI, and H_D are 99.609%, 33.4635%, and 50% respectively. Table 6 shows the results obtained of NPCR, UACI, and H_D for the plain images Lena, Pepper, Baboon, Barbara, and Boats of the same size—256 × 256 grayscale images. As we can see from these results, the NPCR, UACI, and H_D values obtained are very close to the optimal values. These values indicate that the proposed SCbSC system is very sensitive to slight modifications of the secret key.

Table 6. Number of pixel change rate (NPCR), unified average changing intensity (UACI), and H_D values.

Test	Lena	Pepper	Baboon	Barbara	Boats
NPCR %	99.5483	99.5452	99.5788	99.5513	99.5529
UACI %	33.7768	33.6530	33.5595	33.7723	33.6886
H_D	0.5015	0.4991	0.4996	0.5009	0.4996

4.2.2. Statistical Analysis

In order to analyze the resilience of the proposed SCbSC system against most statistical attacks, we use histogram, chi-square, entropy, and correlation analysis.

Histogram and Chi-Square Analysis

The histogram of an encrypted image is an important feature in evaluating the performance of the encryption process. It illustrates how the gray levels of the pixels in an image are distributed and should be very close to a uniform distribution. In Figures 9–13, we give the results obtained for Lena, Peppers, Baboon, Barbara, and Boats of size 256 × 256, in (a) and (c) the plain/cipher images and in (b) and (d) their histograms respectively.

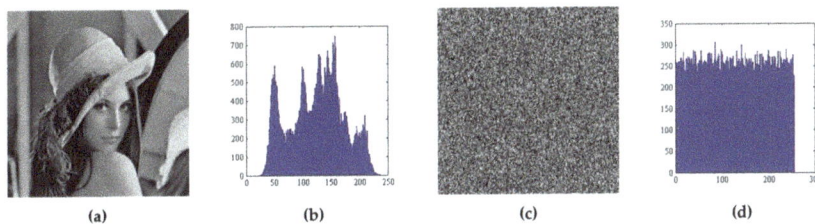

Figure 9. Result of Lena image. (**a**) Lena image, (**b**) histogram of Lena image, (**c**) encrypted Lena, and (**d**) histogram of encrypted Lena.

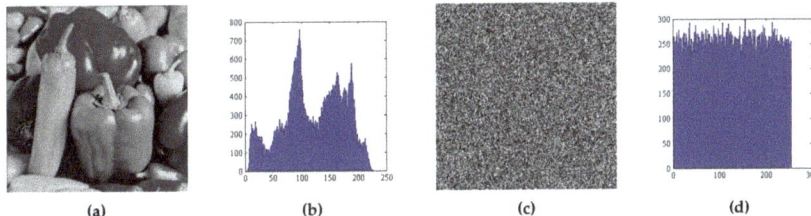

Figure 10. Result of Pepper image. (**a**) Pepper image, (**b**) histogram of Pepper image, (**c**) encrypted Pepper, and (**d**) histogram of encrypted Pepper.

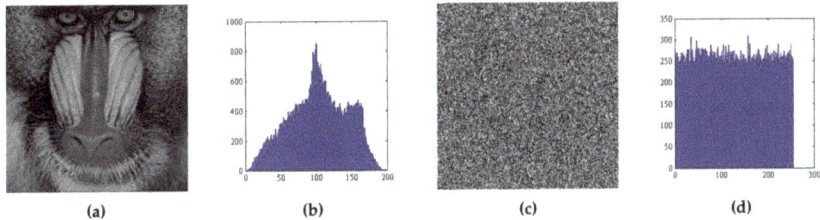

Figure 11. Result of Baboon image. (**a**) Baboon image, (**b**) histogram of Baboon image, (**c**) encrypted Baboon, and (**d**) histogram of encrypted Baboon.

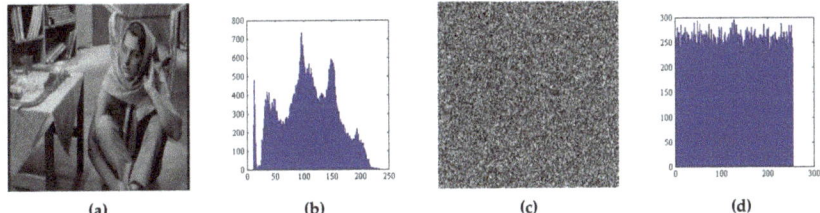

Figure 12. Result of Barbara image. (**a**) Barbara image, (**b**) histogram of Barbara image, (**c**) encrypted Barbara, and (**d**) histogram of encrypted Barbara.

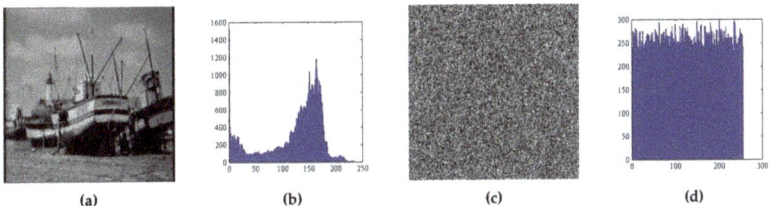

Figure 13. Result of Boats image. (**a**) Boats image, (**b**) histogram of Boats image, (**c**) encrypted Boats, and (**d**) histogram of encrypted Boats.

It was observed that the histograms of the ciphered images are very close to the uniform distribution and are completely different from the plain images. We applied the chi-square test, using Equation (23), on ciphered images to statistically confirm their uniformity. $N_c = 2^8 = 256$ is the number of levels, O_i is the calculated occurrence frequency of each gray level $i \in [0, 255]$ in the histogram of the ciphered image, and E_i is the expected occurrence frequency of the uniform distribution, calculated by $E_i = image\ size\ in\ bytes/N_c$. The distribution of the histogram tested is uniform if it satisfies the following condition: $\chi^2_{ex} < \chi^2_{th}(N_c - 1, \alpha) = 293.24$ (for $N_c = 256$ and $\alpha = 0.05$). The results obtained for the chi-square test, given in Table 7, indicate that the histograms of the ciphered images tested are uniform because their experimental values are smaller than the theoretical values.

Table 7. Chi-square results on the histograms tested.

Chi-Square Test	Lena	Pepper	Baboon	Barbara	Boats
χ^2_{ex}	216.10	231.84	244.05	233.62	265.47
χ^2_{th} (255, 0.05)	293.24	293.24	293.24	293.24	293.24

Entropy Analysis

The random behavior of the ciphered image can be quantitatively measured by entropy information given by Shannon [40]:

$$H(C) = - \sum_{i=0}^{N_c-1} P(c_i) \times log_2(P(c_i)) \quad (29)$$

where $H(C)$ is the entropy of the encrypted image, and $P(c_i)$ is the probability of each gray level appearance ($c_i = 0, 1, \ldots, 255$). In the case of equal probability levels, the entropy is maximum (=8). The closer the experimental entropy value is to the maximum value, the more robust the encryption algorithm. We give in Table 8, the results obtained from the entropy test on the plain and encrypted images. It is clear that the obtained entropies of ciphered images are close to the optimal value. Then, from these results, the proposed stream cipher has a high degree level of resilience.

Table 8. Entropy results obtained.

Entropy	Lena	Pepper	Baboon	Barbara	Boats
Plain image	7.4504	7.5939	7.3102	7.5199	7.2392
Cipher image	7.9571	7.9550	7.9545	7.9570	7.9567

Correlation Analysis

In an original image, each pixel is highly-correlated with adjacent pixels in a horizontal, vertical, and diagonal directions. A good encryption algorithm should produce encrypted images with correlation and redundancy as low as possible (close to zero) between adjacent pixels. To assess the correlation, we performed the following: first, we randomly selected 8000 pairs of two adjacent pixels from the image; then we calculated the correlation coefficients by using the following equation:

$$\rho_{xy} = \frac{Cov(x,y)}{\sqrt{D(x)}\sqrt{D(y)}} \quad (30)$$

where:

$$Cov(x,y) = \frac{1}{N} \sum_{i=1}^{N} [x_i - E(x)][y_i - E(y)] \quad (31)$$

$$E(x) = \frac{1}{N} \sum_{i=1}^{N} x_i \qquad (32)$$

$$D(x) = \frac{1}{N} \sum_{i=1}^{N} [x_i - E(x)]^2 \qquad (33)$$

where x and y are the grayscale values of two adjacent pixels in the image. The obtained results are shown in Table 9.

Table 9. Correlation coefficients of two adjacent pixels in the plain and ciphered images.

Image	Horizontal	Vertical	Diagonal
Lena	0.939403	0.971060	0.931085
Lena encrypted	−0.003684	−0.009015	0.002278
Peppers	0.959869	0.967869	0.940375
Peppers encrypted	−0.005938	−0.004665	−0.001154
Baboon	0.877794	0.834230	0.788141
Baboon encrypted	−0.006750	−0.005998	−0.002088
Barbara	0.907829	0.946119	0.883508
Barbara encrypted	−0.008293	−0.010526	0.004815
Boats	0.940837	0.953357	0.904555
Boats encrypted	−0.002802	−0.009909	0.002302

It appears from Table 9 that the correlation coefficients for the plain images are close to 1, which shows that the pixels are highly correlated, whereas for the encrypted images, the correlation coefficients are close to 0, which proves that there is no correlation between the plain and ciphered images. Therefore, there is no similarity between plain and encrypted images, proving the very good achieved confusion by the proposed SCbSC.

According to all these results of the histogram, entropy, and correlation, the proposed stream cipher presents a good ability to resist statistical attacks.

5. Conclusions

In this paper, we studied and implemented on a Xilinx PYNQ-Z2 FPGA hardware platform using VHDL a novel chaos-based stream cipher (SCbSC) using a proposed secure pseudo-chaotic number generator (SPCNG). The proposed chaotic system includes some countermeasures against side channel attacks (SCAs) and uses a weekly coupling matrix, which prevents division and conquers attacks on the initial vector (IV). Next, we analyzed the cryptographic properties of the proposed SPCNG and evaluated the performances of its hardware metrics. The results obtained demonstrate, on the one hand, the high degree of security, and on the other hand, the good hardware metrics achieved by the SCPNG. After that, we realized the SCbSC system and asserted its resilience against cryptanalytic attacks. Further, we evaluated its hardware metrics and compared them to those of some chaotic and non-chaotic systems. All the results obtained indicate that the proposed SCbSC is a good candidate for encrypting private data. Our future work will focus on designing a chaos-based block cipher to secure IoT data and to check hardware implementations when using non-volatile FPGA technology, which reduces the side attack possibilities in real-field applications.

Author Contributions: Writing—original draft, F.D.; Writing—review & editing, F.D. and S.E.A.; Validation, W.E.H.Y., M.M. and R.L. All authors have read and agreed to the published version of the manuscript.

Funding: This research received no external funding.

Data Availability Statement: Not applicable.

Conflicts of Interest: The authors declare no conflict of interest.

References

1. Lorenz, E.N.; Haman, K. The essence of chaos. *Pure Appl. Geophys.* **1996**, *147*, 598–599.
2. Wang, X.-Y.; Zhang, J.-J.; Zhang, F.-C.; Cao, G.-H. New chaotical image encryption algorithm based on Fisher–Yatess scrambling and DNA coding. *Chin. Phys. B* **2019**, *28*, 040504. [CrossRef]
3. Belazi, A.; Abd El-Latif, A.A.; Belghith, S. A novel image encryption scheme based on substitution-permutation network and chaos. *Signal Process.* **2016**, *128*, 155–170. [CrossRef]
4. Amigo, J.; Kocarev, L.; Szczepanski, J. Theory and practice of chaotic cryptography. *Phys. Lett. A* **2007**, *366*, 211–216. [CrossRef]
5. Kocarev, L. Chaos-based cryptography: A brief overview. *IEEE Circuits Syst. Mag.* **2001**, *1*, 6–21. [CrossRef]
6. Acho, L. A chaotic secure communication system design based on iterative learning control theory. *Appl. Sci.* **2016**, *6*, 311. [CrossRef]
7. Datcu, O.; Macovei, C.; Hobincu, R. Chaos Based Cryptographic Pseudo-Random Number Generator Template with Dynamic State Change. *Appl. Sci.* **2020**, *10*, 451. [CrossRef]
8. Abdoun, N.; El Assad, S.; Manh Hoang, T.; Deforges, O.; Assaf, R.; Khalil, M. Designing Two Secure Keyed Hash Functions Based on Sponge Construction and the Chaotic Neural Network. *Entropy* **2020**, *22*, 1012. [CrossRef] [PubMed]
9. Battikh, D.; El Assad, S.; Hoang, T.M.; Bakhache, B.; Deforges, O.; Khalil, M. Comparative Study of Three Steganographic Methods Using a Chaotic System and Their Universal Steganalysis Based on Three Feature Vectors. *Entropy* **2019**, *21*, 748. [CrossRef]
10. Liao, T.-L.; Wan, P.-Y.; Yan, J.-J. Design of synchronized large-scale chaos random number generators and its application to secure communication. *Appl. Sci.* **2019**, *9*, 185. [CrossRef]
11. Pareek, N.K.; Patidar, V.; Sud, K.K. Image encryption using chaotic logistic map. *Image Vis. Comput.* **2006**, *24*, 926–934. [CrossRef]
12. Kocarev, L.; Jakimoski, G. Logistic map as a block encryption algorithm. *Phys. Lett. A* **2001**, *289*, 199–206. [CrossRef]
13. François, M.; Grosges, T.; Barchiesi, D.; Erra, R. Pseudo-random number generator based on mixing of three chaotic maps. *Commun. Nonlinear Sci. Numer. Simul.* **2014**, *19*, 887–895. [CrossRef]
14. Wang, X.-Y.; Qin, X. A new pseudo-random number generator based on CML and chaotic iteration. *Nonlinear Dyn.* **2012**, *70*, 1589–1592. [CrossRef]
15. Taha, M.A.; Assad, S.E.; Queudet, A.; Deforges, O. Design and efficient implementation of a chaos-based stream cipher. *Int. J. Internet Technol. Secur. Trans.* **2017**, *7*, 89–114. [CrossRef]
16. Jallouli, O.; El Assad, S.; Chetto, M.; Lozi, R. Design and analysis of two stream ciphers based on chaotic coupling and multiplexing techniques. *Multimed. Tools Appl.* **2018**, *77*, 13391–13417. [CrossRef]
17. Lozi, R. Emergence of randomness from chaos. *Int. J. Bifurc. Chaos* **2012**, *22*, 1250021. [CrossRef]
18. Ding, L.; Liu, C.; Zhang, Y.; Ding, Q. A new lightweight stream cipher based on chaos. *Symmetry* **2019**, *11*, 853. [CrossRef]
19. Abdelfatah, R.I.; Nasr, M.E.; Alsharqawy, M.A. Encryption for multimedia based on chaotic map: Several scenarios. *Multimed. Tools Appl.* **2020**. [CrossRef]
20. Gautier, G.; Le Glatin, M.; El Assad, S.; Hamidouche, W.; Déforges, O.; Guilley, S.; Facon, A. Hardware Implementation of Lightweight Chaos-Based Stream Cipher. In Proceedings of International Conference on Cyber-Technologies and Cyber-Systems, Porto, Portugal, 22 September 2019; 5p.
21. Tanougast, C. Hardware implementation of chaos based cipher: Design of embedded systems for security applications. In *Chaos-Based Cryptography*; Springer: Berlin/Heidelberg, Germany, 2011; pp. 297–330.
22. Koyuncu, İ.; Tuna, M.; Pehlivan, İ.; Fidan, C.B.; Alçın, M. Design, FPGA implementation and statistical analysis of chaos-ring based dual entropy core true random number generator. *Analog Integr. Circuits Signal Process.* **2020**, *102*, 445–456. [CrossRef]
23. Nguyen, R. *Penetration Testing on a C-Software Implementation aff1709rns006-c*; Internal Report; Secure-IC SAS: Cesson-Sévigné, France, 2018.
24. Nguyen, R.; Facon, A.; Guilley, S.; Gautier, G.; El Assad, S. Speed-up of SCA Attacks on 32-bit Multiplications. In Proceedings of the International Conference on Codes, Cryptology, and Information Security, Rabat, Morocco, 22–24 April 2019; pp. 31–39.
25. Peng, J.; You, M.; Yang, Z.; Jin, S. Research on a block encryption cipher based on chaotic dynamical system. In Proceedings of the Third International Conference on Natural Computation (ICNC 2007), Haikou, China, 24–27 August 2007; pp. 744–748.
26. Masuda, N.; Jakimoski, G.; Aihara, K.; Kocarev, L. Chaotic block ciphers: From theory to practical algorithms. *IEEE Trans. Circuits Syst. I Regul. Pap.* **2006**, *53*, 1341–1352. [CrossRef]
27. El Assad, S. *Chaos-Based Cryptography, Internal Report*; University of Nantes: Nantes, France, 2019.
28. Jallouli, O. Chaos-Based Security under Real-Time and Eenergy Constraints for the Internet of Things. Ph.D. Thesis, University of Nantes, Nantes, France, 2017.
29. Blackman, D.; Vigna, S. Scrambled linear pseudorandom number generators. *arXiv* **2018**, arXiv:1805.01407.
30. Vigna, S. Further scramblings of Marsaglia's xorshift generators. *J. Comput. Appl. Math.* **2017**, *315*, 175–181. [CrossRef]
31. Coron, J.-S.; Rondepierre, F.; Zeitoun, R. High order masking of look-up tables with common shares. *Iacr Trans. Cryptogr. Hardw. Embed. Syst.* **2018**, 40–72. [CrossRef]
32. Coron, J.-S.; Roy, A.; Vivek, S. Fast evaluation of polynomials over binary finite fields and application to side-channel countermeasures. In *International Workshop on Cryptographic Hardware and Embedded Systems*; Springer: Berlin/Heidelberg, Germany, 2014; pp. 170–187.
33. Rukhin, A.; Soto, J.; Nechvatal, J.; Smid, M.; Barker, E. *A Statistical Test Suite for Random and Pseudorandom Number Generators for Cryptographic Applications*; Booz-allen and Hamilton Inc.: McLean, VA, USA, 2001.

34. Manifavas, C.; Hatzivasilis, G.; Fysarakis, K.; Papaefstathiou, Y. A survey of lightweight stream ciphers for embedded systems. *Secur. Commun. Networks* **2016**, *9*, 1226–1246. [CrossRef]
35. Maximov, A.; Biryukov, A. Two trivial attacks on Trivium. In *International Workshop on Selected Areas in Cryptography*; Springer: Berlin/Heidelberg, Germany, 2007; pp. 36–55.
36. Gaj, K.; Southern, G.; Bachimanchi, R. Comparison of hardware performance of selected Phase II eSTREAM candidates. In Proceedings of the State of the Art of Stream Ciphers Workshop (SASC 2007), eSTREAM, ECRYPT Stream Cipher Project, Report, Lausanne, Switzerland, 31 January–1 February 2007.
37. Bulens, P.; Kalach, K.; Standaert, F.-X.; Quisquater, J.-J. FPGA implementations of eSTREAM phase-2 focus candidates with hardware profile. In Proceedings of the State of the Art of Stream Ciphers Workshop (SASC 2007), eSTREAM, ECRYPT Stream Cipher Project, Report, Lausanne, Switzerland, 31 January–1 February 2007.
38. Schneier, B. *Applied Cryptography: Protocols, Algorithms, and Source Code in C*; John Wiley & Sons: Hoboken, NJ, USA, 2007.
39. Wu, Y.; Noonan, J.P.; Agaian, S. NPCR and UACI randomness tests for image encryption. *CYber J. Multidiscip. J. Sci. Technol. Sel. Areas Telecommun.* **2011**, *1*, 31–38.
40. Wu, Y.; Zhou, Y.; Saveriades, G.; Agaian, S.; Noonan, J.P.; Natarajan, P. Local Shannon entropy measure with statistical tests for image randomness. *Inf. Sci.* **2013**, *222*, 323–342. [CrossRef]

Article

Bit Independence Criterion Extended to Stream Ciphers

Evaristo José Madarro-Capó [1], Carlos Miguel Legón-Pérez [1], Omar Rojas [2], Guillermo Sosa-Gómez [2,*] and Raisa Socorro-Llanes [3]

1. Institute of Cryptography, University of Havana, Havana 10400, Cuba; ejmcapo@gmail.com (E.J.M.-C.); clegon58@gmail.com (C.M.L.-P.)
2. Facultad de Ciencias Económicas y Empresariales, Universidad Panamericana, Álvaro del Portillo 49, Zapopan, Jalisco 45010, Mexico; orojas@up.edu.mx
3. Faculty of Informatics, Technological University of Havana (UTH), CUJAE, Havana 19390, Cuba; raisa@ceis.cujae.edu.cu
* Correspondence: gsosag@up.edu.mx; Tel.: +52-3313682200

Received: 30 September 2020; Accepted: 26 October 2020; Published: 29 October 2020

Abstract: The bit independence criterion was proposed to evaluate the security of the S-boxes used in block ciphers. This paper proposes an algorithm that extends this criterion to evaluate the degree of independence between the bits of inputs and outputs of the stream ciphers. The effectiveness of the algorithm is experimentally confirmed in two scenarios: random outputs independent of the input, in which it does not detect dependence, and in the RC4 ciphers, where it detects significant dependencies related to some known weaknesses. The complexity of the algorithm is estimated based on the number of inputs l, and the dimensions, n and m, of the inputs and outputs, respectively.

Keywords: bit independence criterion; bit independence; RC4; stream cipher; complexity

1. Introduction

Randomness is an essential component in the security of cryptographic algorithms [1,2]. In particular, stream ciphers are composed of pseudo-random number generators and base their security on the statistical characteristics of these generators [1]. Several stream ciphers can be found in the literature whose description is based on different methods for the generation of pseudo-random numbers [3].

In practice, to determine if a generator is suitable to be used for cryptographic purposes, several statistical tests are usually applied on it to measure the randomness of its outputs [4–6]. There are numerous statistical tests to measure the randomness of the outputs of a pseudo-random number generator, among these those grouped in the batteries of NIST [7], Diehard [8], TestU01 [9], and Knuth [10], among others [2]. However, despite a large number of statistical tests being present in these batteries, none of them measure the correlation between the inputs and outputs of the stream cipher; they only measure the randomness of the outputs, which is a necessary, but not sufficient, condition to consider the generator for use in cryptography.

To consider a stream cipher secure, there must be no statistically significant correlation between the structure of its inputs and outputs. If "patterns" depending on the structure of the cipher input are generated in the output of stream ciphers, this could provide information about the input used. In the literature, there are reports of cryptanalysis based on this type of weakness [11,12]. In this way, it is essential to avoid the previous weakness and to have methods to detect it in the design and evaluation stage of the algorithm; in particular, it is necessary to have statistical tests that are capable of detecting the existence of significant statistical dependencies between the inputs and outputs of stream ciphers. In general, there are very few statistical test reports to detect the existence of statistical dependencies between the outputs and inputs of a stream cipher. Therefore, the design of statistical tests that allow for the evaluation of them in this sense is highly important in cryptography.

The strict avalanche criterion (SAC) and the bit independence criterion (BIC) were proposed in [13] to evaluate the strength of the S-boxes used in block ciphers [14]. These two criteria measure different characteristics of the change's effect that an input bit has on the output bits; while the SAC verifies uniformity in the distribution of each output bit, the BIC measures the degree of independence between the output bits [15]. The SAC has been extended to be applied to stream ciphers [16–22]. In [22], the RC4 stream cipher [23] was evaluated through the SAC and the existence of statistical dependence between the input bits and outputs of the RC4 was detected for inputs of large size. This confirms the results obtained in [24–27], where the existence of related inputs in RC4 was reported. The idea developed in [22] was to determine the behavior of the distribution of the bits in the output by changing any bit in the input. In the design of stream ciphers, the distribution behavior of the output elements must be uniformly distributed, regardless of the bit that is being changed at the input [5]. Otherwise, the outputs could provide information on the input bits, which constitutes a weakness that, in the worst-case scenario, could lead to an attack. A discussion of attacks on stream ciphers can be found in [28]. However, the BIC has not been applied, to the best of our knowledge, to assess the degree of statistical independence between the bits of the output stream ciphers from changing a bit of the input. In this paper, we propose an algorithm that extends this criterion to evaluate the degree of independence between the input bits and the outputs of the stream ciphers. The effectiveness of the algorithm was experimentally confirmed in two scenarios: random outputs independent of the input, in which it does not detect dependence, and in the RC4 cipher, where it detects significant dependencies related to some known weaknesses [22,24–26].

2. Preliminaries

A stream cipher can be viewed as a function $f : \mathbb{F}_2^n \to \mathbb{F}_2^m$ that transforms a binary input vector $X = (x_1, \ldots, x_n)$ of n bits into a binary output vector $Y = f(X) = (y_1, \ldots, y_m)$ of m bits, where $n, m \in \mathbb{N}$. In [13], the difference between the outputs $Y = f(X)$ and $Y^i = f(X^i)$, corresponding to the inputs X and X^i, is called the avalanche vector and denoted by $V^i = Y \oplus Y^i$, where $X^i = X \oplus e_i$, with $1 \leq i \leq n$ and e_i the unit vector with 1 in the i-th component. In $V^i = Y \oplus Y^i = (v_1^i, v_2^i, \ldots, v_m^i)$ each $v_j^i \in \mathbb{F}_2$, with $1 \leq j \leq m$, is called an avalanche variable (see Table A1, Appendix A).

Given the set $D = \{X_1, \ldots, X_l\}$ of l inputs X_r of n bits, with $1 \leq r \leq l$, a binary matrix H^i is constructed for each e_i, $1 \leq i \leq n$. To construct the matrix H^i, the avalanche vectors $V_r^i = Y_r \oplus Y_r^i = (v_{r1}^i, v_{r2}^i, \ldots, v_{rm}^i)$ are calculated, with $Y_r = f(X_r)$, $Y_r^i = f(X_r \oplus e_i)$. It is said that f satisfies the BIC if, by changing any bit i in the l inputs $X_r \in D$, it is satisfied that every pair of avalanche variables $v_{\cdot j}^i$ and $v_{\cdot k}^i$ are independent, with $1 \leq j, k \leq m$. The matrix H^i will be called the SAC matrix associated with the vector e_i and is shown in Table 1.

To measure the degree of independence between the pairs of avalanche variables, Webster and Tavares [13] used Pearson's correlation coefficient. In [29], the maximum value of these coefficients was used as a test statistic, denoted here by

$$BIC_{Pearson}(f) = \max_{\substack{1 \leq i \leq n \\ 1 \leq j,k \leq m \\ j \neq k}} \rho(v_{\cdot j}^i, v_{\cdot k}^i). \qquad (1)$$

If all pairs of avalanche variables $v_{\cdot j}^i$ and $v_{\cdot k}^i$ are independent, then ideally, $BIC_{Pearson}(f) = 0$. Therefore, in practice, when $BIC_{Pearson}(f) \approx 0$, it is concluded that f satisfies the BIC.

Table 1. SAC matrix $H^i = (v^i_{rj})$ of dimension $l \times m$ for the change of bit i over the set D of l inputs.

Avalanche Vectors	Avalanche Variables						
	$v^i_{.1}$	$v^i_{.2}$	...	$v^i_{.j}$	...	$v^i_{.k}$	...	$v^i_{.m}$
V^i_1	v^i_{11}	v^i_{12}	...	v^i_{1j}	...	v^i_{1k}	...	v^i_{1m}
$\vdots$	$\vdots$	$\vdots$	$\vdots$	$\vdots$	$\vdots$	$\vdots$	$\vdots$	$\vdots$
V^i_r	v^i_{r1}	v^i_{r2}	...	v^i_{rj}	...	v^i_{rk}	...	v^i_{rm}
$\vdots$	$\vdots$	$\vdots$	$\vdots$	$\vdots$	$\vdots$	$\vdots$	$\vdots$	$\vdots$
V^i_l	v^i_{l1}	v^i_{l2}	...	v^i_{lj}	...	v^i_{lk}	...	v^i_{lm}

2.1. Comparison between SAC and BIC

The SAC [13] verifies whether each output bit changes approximately half of the time by changing an input bit. Using the SAC matrix H^i, it is said that f satisfies the SAC if for all i and every avalanche variable $v^i_{.j}$, with $1 \leq j \leq m$ and $1 \leq i \leq n$, $HW(v^i_{.j})$ is binomial distributed with parameters $n = l$ and $p = \frac{1}{2}$, i.e., $v^i_{.j} \sim B\left(l, \frac{1}{2}\right)$, where $HW(\cdot)$ is the Hamming weight. On the other hand, the BIC [13] measures the degree of independence between each pair $v^i_{.j}, v^i_{.k}$ of avalanche variables. Thus, the two criteria measure a different characteristic from the effect produced on the output bits changing an input bit; the SAC verifies uniformity in the distribution of each output bit, while the BIC measures the degree of independence between the output bits.

In [30], a new method to assess the correlation between statistical randomness tests based on mutual information was presented, using some test statistics and p-values of the tests. This tool can be used to determine the degree of correlation between these two statistical tests. In [29], an assessment of the independence between these two tests through absolute correlation coefficient is given, concluding that these tests are quite uncorrelated.

2.2. Stream Ciphers and RC4

The stream ciphers perform the encryption by converting plain text into bit-by-bit cipher-text through the use of a keystream and the XOR operation. A keystream is nothing more than a sequence of numbers generated in a pseudo-random way. This is achieved by building a pseudo-random number generator. The sequence of pseudo-random numbers used must meet certain statistical properties to be considered suitable for cryptographic use. In many applications (see [4,31]), ciphers of this type have become very important tools since they are very fast and their implementation is simpler than other ciphers, e.g., a block cipher. In these types of scenarios, the problem is in the transmission of a large amount of data in communication networks in a short time.

There are a wide variety of design proposals [32] to build pseudo-random number generators. Among these, the RC4 algorithm [23] stands out from others for its wide use in different applications and protocols. The RC4 stream cipher [23] is optimized to be used in 8-bit processors, being extremely fast and exceptionally simple. It was included in network protocols such as Secure Sockets Layer (SSL), Transport Layer Security (TLS), Wired Equivalent Privacy (WEP), Wi-Fi Protected Access (WPA), and in various applications used in Microsoft Windows, Lotus Notes, Apple Open Collaboration Environment (AOCE), and Oracle Secure SQL [23]. In the last decade, some applications [33,34] avoided RC4 encryption given some weaknesses found [35]. However, although it is not considered very secure [36], RC4 is still one of the most widely used stream ciphers [37], and continues to motivate research nowadays [36–38]. Furthermore, this cipher is a good option to measure the effectiveness of methods that analyze weaknesses in stream ciphers related to those already known in RC4 [22,24–26], or to check the performance of hardware or software schemes that make use of cryptography [39–41].

The RC4 has two main components: the key scheduling, and the pseudo-random number generator. The key scheduling generates an internal random permutation S of values from 0 to 255, from an initial permutation, a (random) key K of l-byte length, and two pointers i and j. The maximal key length is of $l = 256$ bytes (see Algorithm 1).

Algorithm 1 RC4 key-scheduling

1: **for** $i = 0 \to 255$ **do**
2:     $S[i] \leftarrow i$
3: **end for**
4: $j \leftarrow 0$
5: **for** $i = 0 \to 255$ **do**
6:     $j \leftarrow (j + S[i] + K[i \mod l]) \mod n$
7:     Swap $S[i]$ and $S[j]$
8: **end for**

The main part of the algorithm is the pseudo-random number generator that produces one-byte output in each step. As usual, for stream ciphers, the encryption will be an XOR of the pseudo-random sequence with the message (see Algorithm 2).

Algorithm 2 RC4 pseudo-random generator

1: $i \leftarrow 0$
2: $j \leftarrow 0$
3: **while** Generating Output **do**
4:     $i \leftarrow (i + 1) \mod 256$
5:     $j \leftarrow (j + S[i]) \mod 256$
6:     Swap $S[i]$ and $S[j]$
7:     Output $S[(S[i] + S[j]) \mod 256]$
8: **end while**

The weaknesses found can be classified according to the theme they exploit, some of which are:

1. Weak keys.
2. Key recovery from the state.
3. Key recovery from the key-stream.
4. State recovery attacks.
5. Biases and distinguishes.

While the fifth point is the most studied subject in the literature, the third point is the most serious attack made to RC4. The theme that is exploited in this paper has been deeply studied—in particular, Grosul and Wallach [24] demonstrated that certain related key-pairs generate similar output bytes in RC4. Later, Matsui [25] reported colliding key pairs for RC4 for the first time, and then stronger key collisions were found in [26]. For the RC4 stream cipher, several modifications have been proposed; while some modified only certain components or some operations, others completely changed the algorithm (see [42]). It is important to note that even RC4 variants have had a lot of attention in the scientific community (see [43]).

3. BIC Algorithm in Stream Ciphers

In this section, an algorithm is proposed to extend the bit independence criterion (BIC) to stream ciphers, experimentally confirming its effectiveness. The two main differences that arise in this scenario with respect to its application in S-boxes are discussed.

Let f be the function that will be evaluated by the BIC, $D = \{X_1, \ldots, X_l\}$ the set of l inputs X_r of n bits generated randomly and m the number of bits of the outputs of f, the proposed method consists of the following steps:

Step 1. Construct the n SAC H^i, $(i = 1, \ldots, n)$ matrices of dimension $l \times m$.

1. Evaluate $Y_r = f(X_r)$, $(r = 1, \ldots, l)$, and generate the output Y_r of size m.
2. Evaluate $Y_r^i = f(X_r \oplus e_i)$, and generate the output Y_r^i of size m, where e_i is the canonical vector.
3. Build the avalanche vector $V_r^i = Y_r \oplus Y_r^i = \{v_{r1}^i, \ldots, v_{rm}^i\}$ of m avalanche variables $v_{\cdot j}^i$, $(j = 1, \ldots, m)$.

Step 2. Evaluate the independence between the avalanche variables $v_{\cdot j}^i$ and $v_{\cdot k}^i$.

1. For each pair (j, k), with $1 \leq j, k \leq m$ and $j \neq k$, measure the independence between the avalanche variables $v_{\cdot j}^i, v_{\cdot k}^i$ by a test statistic.
2. Set a significance level α_1 and decide, using a statistical criterion, if the observed value of the test statistic allows to reject or not the hypothesis of independence between $v_{\cdot j}^i$ and $v_{\cdot k}^i$.
3. Count the number T^i of rejections between C_2^m pairs of the matrix H^i.

Step 3. Decision on whether or not to comply with the BIC criterion:

1. Count the total number T of rejections between the n matrices H^i.
2. Set a significance level α_2.
3. Decide, using a statistical criterion, whether the observed value of T allows to reject the BIC compliance.

The following sections describe each of these steps and end with the proposal of an algorithm to evaluate the BIC in stream ciphers.

3.1. Building the SAC Matrix

First difference. When evaluating the BIC in S-boxes, it is possible to go through the entire space of $l = 2^n$ inputs since n usually takes small values; however, this is impractical in stream ciphers where the dimension of the input space can be 2^{128} or greater. To solve this problem, it is proposed to use the same approach applied in the randomness assessment to the outputs of pseudo-random generators through statistical tests [2]. This approach consists of generating a sample of l inputs with $l \ll 2^n$, and to determine the strength of the cipher from the results obtained from this sample.

The l inputs are chosen randomly in the space of 2^n possible inputs. This is the main difference; while the BIC test works over all of the input space with S-boxes, the stream cipher works with a randomly selected subset of the sample space.

3.2. Test of Independence between Two Avalanche Variables $v_{\cdot j}^i$ and $v_{\cdot k}^i$

Second difference. In [13], Pearson's correlation coefficient ρ was used to measure the degree of independence between the pairs of avalanche variables. The use of such a coefficient in [13,29] has two main disadvantages: the first one is that it only detects linear correlations, and the second one is that the critical region for the rejection of the null hypothesis is not explicitly defined, i.e., a threshold is not defined below which $BIC_{Pearson}(f) \approx 0$ is decided. Thus, it can be a reason for an imprecision in the decision when dealing with small coefficient values. In order to solve the first aforementioned disadvantage, mutual information can be applied to measure the degree of independence between pairs of avalanche variables [44], but in this case, it is important to determine which estimator to use, since there are no estimators of unbiased entropy of minimal variance; the second disadvantage can be solved by defining the critical region using a transformation of the correlation coefficient of the type $t = \sqrt{(N-2)\rho^2/(1-\rho^2)}$, where t is distributed as a t-Student distribution with $N-2$ degrees of freedom [45].

Another approach is that when $v^i_{\cdot j}$ and $v^i_{\cdot k}$ are independent, then $s^i_{jk} = v^i_{\cdot j} \oplus v^i_{\cdot k}$ is balanced [46]. In this work, independence will be evaluated by measuring the adjustment $HW(s^i_{jk})$ to the binomial distribution $B(l, 1/2)$, where $HW(\cdot)$ is the Hamming weight. This allows setting a threshold for the decision criterion on independence between $v^i_{\cdot j}$ and $v^i_{\cdot k}$.

Since H^i is a binary matrix, the adjustment to the binomial distribution will be measured by the χ^2-test with 1 degree of freedom, with the test hypothesis given by:

$$H_0 : v^i_{\cdot j} \text{ and } v^i_{\cdot k} \text{ independent,}$$
$$H_1 : v^i_{\cdot j} \text{ and } v^i_{\cdot k} \text{ dependent.}$$

That is,

$$H_0 : HW(s^i_{jk}) \sim B\left(l, \frac{1}{2}\right),$$
$$H_1 : HW(s^i_{jk}) \not\sim B\left(l, \frac{1}{2}\right).$$

The test statistic used is

$$\chi^2_{s^i_{jk}} = \frac{\left(HW(s^i_{jk}) - \frac{l}{2}\right)^2}{\frac{l}{4}}. \tag{2}$$

As usual [2], the value α_1 is such that

$$P\left(\chi^2_{s^i_{jk}} \leq \chi^2_{\alpha_1, 1}\right) = 1 - \alpha_1. \tag{3}$$

If $\chi^2_{s^i_{jk}} > \chi^2_{\alpha_1, 1}$ the null hypothesis H_0 is rejected.

It is left for future works, to compare the effectiveness of these three criteria for evaluating independence between the avalanche variables.

3.3. BIC Acceptance Test

To decide whether the stream cipher f satisfies the BIC, it is necessary to take into account the number of rejections of H_0 on the n matrices; for this, a random variable T, which counts the total number of rejections on n matrices is defined:

$$T = T(n, m, \alpha_1) = \sum_{i=1}^{n} T^i(m, \alpha_1), \tag{4}$$

where

$$T^i(m, \alpha_1) = T^i = \sum_{j=1}^{m-1} \sum_{k>j}^{m} t\left(v^i_{\cdot j}, v^i_{\cdot k}, \alpha_1\right), \tag{5}$$

and

$$t(v^i_{\cdot j}, v^i_{\cdot k}, \alpha_1) = \begin{cases} 1 & \text{If } H_0 \text{ is rejected for } v^i_{\cdot j} \text{ and } v^i_{\cdot k} \\ & \text{with significance } \alpha_1 \\ 0 & \text{otherwise.} \end{cases} \tag{6}$$

The variable T^i counts the number of rejections of the null hypothesis H_0 in the matrix H^i.

Expected number of rejections of H_0. In each of the n SAC H^i matrices, C_2^m pairs of columns are formed, thus the number of rejections T satisfies

$$0 \leq T \leq n \cdot C_2^m. \tag{7}$$

When $T = 0$, we have the ideal case for compliance with the BIC, since all the pairs of columns are independent, while as $T \gg 0$, the number of non-independent column pairs increases.

Under the hypothesis test above, with a significance level α_1, the expected number of rejections of H_0 is:

$$E\left(T^i | H_0\right) = (\alpha_1 \cdot C_2^m), \tag{8}$$

for each matrix H^i. In total, among the n matrices SAC are expected

$$E(T | H_0) = n \cdot (\alpha_1 \cdot C_2^m), \tag{9}$$

H_0 rejections.

The random variable

$$T = \sum_{i=1}^{n} \sum_{j=1}^{m-1} \sum_{k>j}^{m} t\left(v_{\cdot j}^i, v_{\cdot k}^i, \alpha_1\right), \tag{10}$$

follows a binomial distribution $B(n \cdot C_2^m, \alpha_1)$. Taking into account that generally $\alpha_1 < 0.1$, this distribution can be approximated, in this case, to the Poisson distribution with parameter $\lambda = (\alpha_1 \cdot n \cdot C_2^m)$. Since λ is large, due to large values of $n \cdot C_2^m$, then the Poisson distribution can be approximated by the Normal distribution with mean and variance:

$$E(T|H_0) = \alpha_1 \cdot n \cdot C_2^m, \quad \sigma^2(T|H_0) = \alpha_1 \cdot n \cdot C_2^m \cdot (1 - \alpha_1). \tag{11}$$

Thus

$$Z_T = \frac{T - E(T|H_0)}{\sqrt{\sigma^2(T|H_0)}} \sim N(0,1). \tag{12}$$

Decision criteria. To compare the Z_T value with the $N(0,1)$ distribution, a significance level α_2 is selected. Then, it is tested if f does not satisfy the BIC, with a significance level α_2, if $Z_T > Z_{1-\alpha_2}$. It can be seen that if $0 \leq T \leq E(T|H_0)$, then the values of Z_T decreases with respect to $Z_{1-\alpha_2}$ and $Z_T > Z_{1-\alpha_2}$ is not satisfied, so the BIC is fulfilled. On the other hand, if $T \gg E(T|H_0)$, then the values of Z_T will be greater as T increases, so $Z_T > Z_{1-\alpha_2}$ is satisfied and the BIC compliance is rejected.

Normality of the test statistic T. In the expression of T there are $n \cdot C_2^m$ Bernoulli variables $t(v_{\cdot j}^i, v_{\cdot k}^i, \alpha_1)$, whose distributions under H_0 and H_1 are different:

Under H_0, all variables $t(v_{\cdot j}^i, v_{\cdot k}^i, \alpha_1)$ are independent, identically distributed and take the value of 1 with probability $p_{jk}^i = P(t(v_{\cdot j}^i, v_{\cdot k}^i, \alpha_1) = 1) = \alpha_1$, so T follows exactly a binomial distribution $B(n \cdot C_2^m, \alpha_1)$. Although generally $\alpha_1 \leq 0.1$ the binomial distribution $B(n \cdot C_2^m, \alpha_1)$ can be approximated by the normal distribution, with mean $E(T|H_0) = \alpha_1 \cdot n \cdot C_2^m$ and variance $\sigma^2(T|H_0) = \alpha_1 \cdot n \cdot C_2^m (1 - \alpha_1)$, taking into account that $n \cdot C_2^m$ grows very quickly with m.

Under H_1, the variables $t(v_{\cdot j}^i, v_{\cdot k}^i, \alpha_1)$ that appear in the expression of T are not identically distributed, since the rejection of the BIC means that there are several matrices H^i for which the hypothesis H_0 of independence between $v_{\cdot j}^i$ and $v_{\cdot k}^i$ is rejected. In this case, $p_{jk}^i \neq \alpha_1$ and may be different when i, j, k varies. For this reason, a binomial does not appear directly as the distribution of T. However, it is still possible to approximate the distribution of T by the Normal distribution. For this it is sufficient to calculate the mean

$$P_{n \cdot C_2^m} = \frac{\sum_{i}^{n} \sum_{j}^{m-1} \sum_{k>j}^{m} p_{jk}^i}{n \cdot C_2^m}, \tag{13}$$

between the probabilities of all the variables $t(v_{\cdot j}^i, v_{\cdot k}^i, \alpha_1)$ and the distribution of T can be approximated by the binomial distribution $B(n \cdot C_2^m, P_{n \cdot C_2^m})$. This distribution, in turn, can be approximated by the Normal distribution, taking into account high values of $n \cdot C_2^m$. The precision of this approximation depends on the difference between the probabilities p^i_{jk} involved in $P_{n \cdot C^m_2}$, therefore the variance value between these probabilities can be a measure of the quality of the approximation.

When comparing the distribution of T under H_0 and H_1, similarities and differences are observed. They are similar in that in both cases T follows a Normal distribution, but there are two differences, the first and most important is observed between the expected values of both distributions (it will be higher under H_1) and the second refers to the level of adjustment to this distribution (may be lower under H_1). In the rest of this work, the proposed method to evaluate the BIC in stream ciphers will be called the BIC test.

3.4. BIC Test Algorithm

Given a set $D = \{X_1, \ldots, X_l\}$ of l randomly chosen n bits inputs to the function f, constructs for each binary vector e_i ($1 \leq i \leq n$) its associated SAC matrix H^i and for all for j, k with $j \neq k$, it is checked if $HW(s^i_{jk})$ follow the $B\left(l, \frac{1}{2}\right)$ distribution, see the proposed Algorithm 3.

Algorithm 3 BIC stream ciphers algorithm

Input: f function to evaluate, n size of the inputs of f, m size of the outputs of f, α_1 and α_2 levels of significance, D set of l inputs to the function f.
Output: If f satisfies the BIC
1: $T = 0$
2: **for** $i = 1 \to n$ **do**
3:     **for** $r = 1 \to l$ **do**   ▷ Matrix Construction H_i
4:         Compute $V^i_r = Y_r \oplus Y^i_r$
5:     **end for**
6:     **for each** (j, k) **do**   ▷ Independence check between $v^i_{\cdot j}$ and $v^i_{\cdot k}$
7:         **if** $\chi^2_{s^i_{jk}} > \chi^2_{\alpha_1, 1}$ **then**
8:             $T = T + 1$ ▷ Independence is rejected between $v^i_{\cdot j}$ and $v^i_{\cdot k}$
9:         **end if**
10:    **end for**
11: **end for**
12: **if** $Z_T > Z_{1-\alpha_2}$ **then** f does not satisfy the BIC
13: **else** f satisfies the BIC
14: **end if**

3.4.1. Complexity of the Algorithm

In steps 3–5 of the algorithm, f is used to generate m output bits. Assuming that the stream cipher f generates each output with a constant cost, then $O(lm)$ operations are performed in these steps, since l times m output bits are generated from f. In steps 6–10 of the algorithm, $O(m^2 l)$ operations are performed due to the computation C^m_2 times the Hamming weight in a sequence of l bits.

Thus the algorithm performs $O\left(n \max(l\, m,\, l\, m^2)\right) = O\left(n\, l\, m^2\right)$ operations, and the number of algorithm operations depends on the number n of input bits, the number m of output bits, and the number l of inputs used. It can be seen that the increase in the parameter m has a greater influence than n and l in increasing the number of operations of the algorithm. In the particular case $m = n = l$, $O(m^4)$ operations are performed.

3.4.2. Parameter Selection

As seen in the previous section, the number of operations of the BIC algorithm depends on three parameters, the number l of inputs, the number n of bits of each input, and the number m of bits of each output.

Selection of l such that $\hat{p} \approx 0.5$ and $HW(s^i_{jk})$ fit to the binomial distribution $B(n, 1/2)$. The number l of entries influences the effectiveness of the χ^2-test in determining whether two columns are independent. Increasing l guarantees a greater fit of $HW(s^i_{jk})$ to the binomial distribution $B(n, 1/2)$; however, it causes an increase in the number of operations. In practice, the idea is to obtain a cost-effectiveness ratio using a value of l such that it maintains the fit and provides a practical number of operations. Using the confidence interval for proportions [47], it is possible to obtain a value of l_0, such that prefixing $l > l_0$ achieves a good fit. This confidence interval is given by

$$P\left(-Z_{\alpha_1/2} < \frac{\hat{p} - p}{\sqrt{\frac{pq}{l}}} < Z_{\alpha_1/2}\right) = 1 - \alpha_1. \tag{14}$$

Solving for l we get to

$$l > l_0 = \frac{Z^2_{\alpha_1/2}\, pq}{e^2}, \tag{15}$$

where $e = \hat{p} - p$, is the deviation of $\hat{p}$ over p, and $q = (1 - p)$.

Example 1. *Calculation of the lower bound l_0 for l. A value l_0 from which, with high probability, it is satisfied that $\hat{q} \approx \hat{p} \approx 0.5$ is needed. Then, substituting for a significance level $\alpha_1 = 0.01$ and a deviation e whose absolute value $|e|$ satisfy inequality $|e| = |\hat{p} - 0.5| \leq 0.03$, we get*

$$l_0 = \frac{Z^2_{0.005} \cdot 0.25}{0.03^2} \approx 2189.$$

In this way, for the significance level α_1 and the deviation e selected, it is concluded that l must be chosen such that $l > l_0 = 2189$.

Example 2. ***Convergence of $\hat{p}$ and deviation $\hat{e}$.*** *Table 2 shows the behavior of the deviation $\hat{e}$ observed for several l, $l > l_0 = 2189$, with $n = 64$ and $m = 32$. It can be seen how, for most of the estimated e, the imposed condition is met $|\hat{e}| \leq 0.03$.*

Table 2. Values of the deviation $|\hat{e}|$ for several l, $l > l_0 = 2189$ with $n = 64$ and $m = 32$.

l	Mean Value $\hat{p}$	$|\hat{e}|$
4096	0.5	0.05
8192	0.5	0.03
16,384	0.5	0.03
32,768	0.5	0.02

Selection of n, m under the null hypothesis H_0. The number n of inputs and the number m of outputs influence the sample size for the calculation of the number T of rejections of H_0. In general, we will have $d = n \cdot C_2^m$ pairs of columns to check and it is expected, with probability α_1, that $\lambda = \alpha_1 \cdot d$ pairs of columns will be rejected.

Let $\lambda_0 = \alpha_1 \cdot d_0$ be some default value of λ from which the distribution of T can be approximated to $N(0, 1)$. It is necessary to select n and m such that $d > d_0$ is satisfied and a value of λ such that $\lambda > \lambda_0$ is obtained. It is advisable to select a high value of λ_0 that avoids the use of corrections and provides a good fit.

It is known that increasing λ_0 provides better precision in the Poisson approximation to the Normal distribution. To obtain d_0, we can use the confidence interval for proportions [47], this time in an approximation to the Normal distribution with one tail. So, we have

$$P\left(\frac{\hat{p} - p}{\sqrt{\frac{pq}{d}}} < Z_{\alpha_2}\right) = 1 - \alpha_2. \tag{16}$$

Solving for d we get to

$$d > d_0 = \frac{Z_{\alpha_2}^2 pq}{e^2}. \tag{17}$$

Example 3. *Calculation of the lower bound d_0 for d. Substituting, $p = 0.01$, $q = 0.99$, with a significance level $\alpha_2 = 0.001$ and a deviation $|e|$ of 0.003, we obtain*

$$d_0 = \frac{Z_{0.001}^2 \cdot 0.25}{0.003^2} \approx 10503.$$

Then, $\lambda_0 = \alpha_1 \cdot d_0 \approx 0.01 \cdot (10503) \approx 105$, therefore, for the values of α_1 and e chosen, it is enough to select values of n and m such that $\lambda > \lambda_0 \approx 105$. In Table 3, for $\alpha_1 = 0.01$, some values of n and m are highlighted in italics from which $\lambda > \lambda_0 = 105$.

Table 3. λ values for multiple values of n and m with $\alpha_1 = 0.01$. Values of n and m are highlighted in italics from which $\lambda > \lambda_0 = 105$.

n	m			
	8	16	32	64
8	2.24	9.6	39.68	161.28
16	4.48	19.2	79.36	322.56
32	8.96	38.4	158.72	645.12
64	17.92	76.8	317.44	1290.24
128	35.84	153.6	634.88	2580.48
256	71.68	307.2	1269.76	5160.96
512	143.36	614.4	2539.52	10,321.92

To select n, m and l, the trade-off between reducing computational cost and maximizing effectiveness can be taken into account. However, it is very important to be careful when selecting which values to use, since minimizing computational cost could limit the effectiveness of the BIC method and overestimate the quality of the stream cipher. It is advised to prioritize increasing effectiveness.

4. Experiments and Discussion of the Results

In this section, experiments are carried out in two different scenarios. In the first scenario, the behavior of the Z_T test statistic is investigated under the hypothesis H_0 of compliance with the BIC test, evaluating the test on random H^i matrices. The second scenario shows the behavior of the Z_T test statistic when evaluating it in a stream cipher that does not meet this criterion.

4.1. Scenario 1 (BIC in Random SAC Matrices)

It is expected that under H_0, we obtain $E(Z_T|H_0) = 0$, $\sigma^2(T|H_0) = 1$ and $Z_T \sim N(0,1)$. The experiments in this scenario were carried out under uniform and independent randomly generated SAC matrices, to evaluate compliance, under H_0, of the $N(0,1)$ distribution of Z_T.

Taking into account Table 3, four sets of parameters were selected, two for $n = m$ and two for $n \neq m$:

- $n = m$: ($n = m = 32$) and ($n = m = 64$)
- $n \neq m$: ($n = 64, m = 32$) and ($n = 8, m = 64$).

Appl. Sci. 2020, *10*, 7668

The values $l \in \{4096, 8192, 16{,}384, 32{,}678\}$ will be varied, in order to verify the influence of the variation of the parameters n, m and l in the adjustment of Z_T. The values of n and m with the lowest computational cost were selected, that is, the values of n and m that provide the lowest values of λ such that $\lambda > \lambda_0 = 105$.

The values n and m will be used as a power of two, since current ciphers work with inputs and outputs whose size has these characteristics and also l to speed up, in terms of execution time, the computation of the BIC method. However, it is important to note that the BIC method can be used for any value of n, m and l, as long as the requirements outlined in the previous section are met.

Normality of Z_T in H^i random matrices. Figure 1 corresponds to the observed distribution of 1000 values of Z_T, for each pair of parameters n and m, and each value of l.

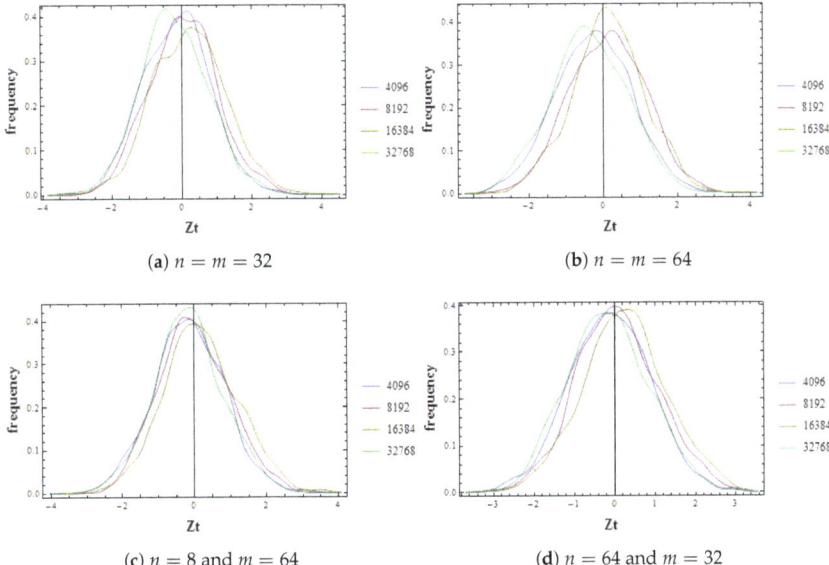

Figure 1. Observed distribution of 1000 values of Z_T in random H^i matrices for various values of n, m, and l.

Tables 4 and 5 show the values $\widehat{E}(Z_T|H_0)$ and $\widehat{\sigma^2}(Z_T|H_0)$ respectively observed in each sample, for each value of n, m and l.

Table 4. Observed $\widehat{E}(Z_T|H_0)$ values for each selected n, m, l value.

(n, m)	l			
	4096	8192	16,384	32,768
(32, 32)	−0.150216	0.044674	0.214355	−0.210047
(64, 64)	−0.268244	0.110717	0.137298	−0.383549
(8, 64)	−0.154163	−0.0239	0.173869	−0.164926
(64, 32)	−0.118008	0.05765	0.236807	−0.175659

The analysis of Figure 1 and Tables 4 and 5, suggests the fulfillment of the hypothesis H_0 about the distribution of $Z_T \sim N(0,1)$, for all the values of the parameters l, n, m selected. As can be seen in Tables 4 and 5, by varying l, n, m, the values $\widehat{E}(Z_T|H_0)$ and $\widehat{\sigma^2}(Z_T|H_0)$ of the observed distribution of

Z_T maintain the fit to the parameters $\mu = 0$ and $\sigma^2 = 1$ expected in a distribution $N(0,1)$. Figure 1 shows the bell shape and approximate symmetry of the obtained distributions.

Table 5. Observed $\widehat{\sigma^2}(Z_T|H_0)$ values for each selected n, m, l value.

(n, m)	l			
	4096	8192	16,384	32,768
(32, 32)	0.906793	0.938818	1.06287	0.930434
(64, 64)	1.04064	1.02569	0.97230	1.04773
(8, 64)	0.972279	0.939652	1.05472	0.923853
(64, 32)	0.994157	0.997373	1.06138	0.983795

Normality Test. The Shapiro–Wilks [48] test for normality was applied to all selected parameter sets. The results are shown in Figure 2 and Table 6.

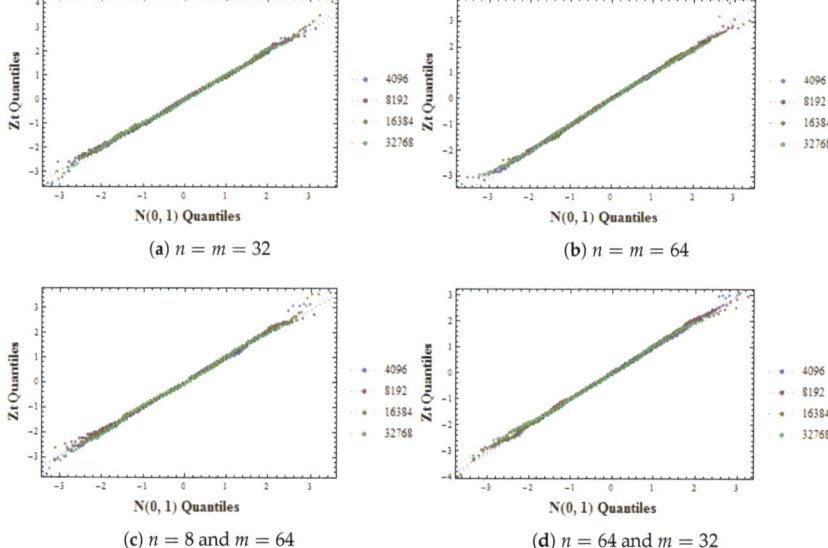

Figure 2. Adjustment of the observed distribution from Z_T to $N(0,1)$ in H^i random matrices, which satisfy the BIC.

In Figure 2 we can see how the observed distribution of Z_t for all the values of l, n, m, fit the distribution $N(0,1)$. Table 6 shows the p-values corresponding to the Shapiro-Wilk normality test for each of the chosen parameter sets.

Table 6. p-values of the Shapiro-Wilk test of normality for samples of Z_t, in random H^i matrices, that satisfy the BIC.

l	n = m = 32	n = m = 64	n = 8 and m = 64	n = 64 and m = 32
4096	0.252382	0.724504	0.262997	0.482318
8192	0.127693	0.573267	0.161048	0.326505
16,384	0.296125	0.653315	0.141577	0.524475
32,768	0.309173	0.37739	0.210961	0.237133

It is observed that in all cases, the p-values are greater than the usual values assumed for α, such as 0.01 or 0.05 and are consistent with the assumed normality hypothesis. The higher the value of $n = m$, the higher the p-value, which corresponds to the influence of these parameters on the value of λ (see Table 3).

BIC test application on H^i random matrices. To evaluate the behavior of the BIC test in random matrices, each Z_t was compared with the critical value $Z_{1-\alpha_2}$, and the number of rejections of H_0 was counted. Tables 7 and 8 show the results for various levels of significance α_2 and $l = 16{,}384$. The observed number of rejections is expected to correspond to that expected according to the selected α_2 level, which would allow choosing α_2, to obtain zero rejections in this scenario.

Table 7. Expected $E(\#\,[Z_T > Z_{1-\alpha_2}|H_0])$ and observed $\#\,[Z_T > Z_{1-\alpha_2}|H_0]$ number of rejections in samples of 1000 values of Z_t for $n = m$, in H^i random matrices.

α_2	$E(\#\,[Z_T > Z_{1-\alpha_2}|H_0])$	$\#\,[Z_T > Z_{1-\alpha_2}|H_0]$	
		$n = m = 32$	$n = m = 64$
0.05	50	31	43
0.01	10	8	9
0.001	1	1	1
0.0001	0	0	0

Table 8. Expected $E(\#\,[Z_T > Z_{1-\alpha_2}|H_0])$ and observed $\#\,[Z_T > Z_{1-\alpha_2}|H_0]$ number of rejections in samples of 1000 values of Z_t for $n \neq m$, in H^i random matrices.

α_2	$E(\#\,[Z_T > Z_{1-\alpha_2}|H_0])$	$\#\,[Z_T > Z_{1-\alpha_2}|H_0]$	
		$n = 8$ and $m = 64$	$n = 64$ and $m = 32$
0.05	50	36	42
0.01	10	7	5
0.001	1	0	1
0.0001	0	0	0

For the value of $\alpha_2 = 0.0001$ located in the last row of both tables, no statistical dependence is detected as expected in random matrices, confirming the effectiveness of the criterion and illustrating the importance of the proper selection of α_2, according to the number $d = n \cdot C_2^m$ of pairs of columns whose independence is evaluated. For the values of $l, n, m, \alpha_1, \alpha_2$ used, such that no Type I error is made, the probability of making a Type II error must be calculated and the values that minimize it must be chosen. In this sense, experiments will be carried out in the second scenario on a stream cipher.

4.2. Scenario 2 (BIC in Stream Cipher)

For this scenario, it is convenient to apply the test to a stream cipher that violates the BIC. RC4 was chosen because there are reports of the existence of dependencies between the inputs and outputs in this cipher [22–25]. Experiments were performed setting the parameters $n = m \in \{32, 64, 128, 160, 256\}$ and 1000 sets D of $l = 16{,}384$ entries each were built. In each set, Z_T was calculated and compared with the critical value $Z_{1-\alpha_2}$, varying α_2. Figure 3 shows the distribution of the 1000 values of Z_T obtained. Table 9 show the values $\widehat{E}(Z_T)$ and $\widehat{\sigma}^2(Z_T)$ observed in each sample, for each value of $n, m,$ and l.

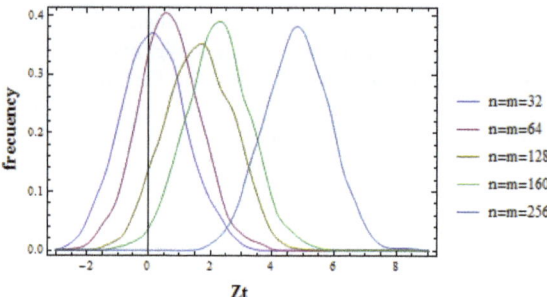

Figure 3. Distribution of the sample of 1000 values of Z_T for SAC matrices generated with RC4 with $n = m \in \{32, 64, 128, 160, 256\}$.

Table 9. Expected value $\widehat{E}(Z_T)$ and variance $\widehat{\sigma^2}(Z_T)$ of Z_T for SAC matrices generated with the RC4.

(n, m)	$\widehat{E}(Z_T)$	$\widehat{\sigma^2}(Z_T)$
(32, 32)	0.149419	1.06442
(64, 64)	0.661726	0.967951
(128, 128)	1.62968	1.15061
(160, 160)	2.24493	1.07417
(256, 256)	4.79748	1.06715

To verify the normality of the data, the Shapiro–Wilks [48] normality test was applied to all the selected parameter sets. The results are shown in Figure 4 and Table 10.

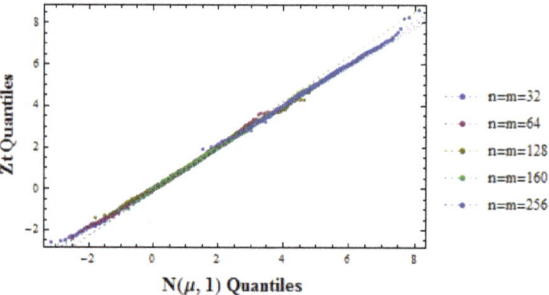

Figure 4. Normality test of the sample of 1000 values of Z_T for SAC matrices generated with the RC4 with $n = m \in \{32, 64, 128, 160, 256\}$.

Table 10. p-values of the Shapiro-Wilk test of normality on samples of Z_t for SAC matrices generated with the RC4 with $n = m \in \{32, 64, 128, 160, 256\}$.

(n, m)	p-Values
(32, 32)	0.103538
(64, 64)	0.582878
(128, 128)	0.382171
(160, 160)	0.943337
(256, 256)	0.673625

In Figure 4 we can see how by increasing the values of $m = n$ the Normal distribution $N(\mu, 1)$ of the statistician Z_t is maintained, however, the value of μ increases (see Figure 3 and Table 9).

It is observed that in all cases the p-values are greater than the usual values assumed for α, such as 0.01 or 0.05 and the samples maintain normality.

In Table 11 it is noted how in RC4 the effectiveness of the criterion increases as the values of n and m increase. That is, increasing the values $m = n$ increases the number of correct decisions to reject H_0. As mentioned, it is known that by increasing the value of n in RC4 the probability of finding very similar outputs, or even the same, increases for inputs that differ by a few bits [22,24–26].

Table 11. Expected $E(\#[Z_T > Z_{1-\alpha_2} \mid H_0])$ and observed $\#[Z_T > Z_{1-\alpha_2}]$ number of rejections in 1000 repetitions of the BIC test in SAC matrices generated with the RC4. All cases in which the observed number of rejections exceeds the expected value are indicated in italics.

α_2	$E(\#[Z_T > Z_{1-\alpha_2} \mid H_0])$	$\#[Z_T > Z_{1-\alpha_2}]$ $n = m$				
		32	64	128	160	256
0.05	50	77	155	497	731	1000
0.01	10	22	44	272	462	993
0.001	1	2	11	91	215	953
0.0001	0	0	0	19	74	843

This experiment confirms the effectiveness of the BIC test by detecting dependence between the inputs-outputs of RC4 and allows us to conclude that in RC4, the effectiveness is an increasing function of the value of the parameters $n = m$. All cases in which the observed number of rejections exceeds the expected value are indicated in italics.

An important feature in statistical tests is the determination of type I and type II errors [2]. Under H_0, we have that $v^i_{.j}$ and $v^i_{.k}$ are independent, then the type I error consists in rejecting independence when they are and therefore deciding that the cipher has a weakness when it does not have it. Meanwhile, not rejecting H_0 when there is a dependency means that it would be decided that the cipher passes the BIC, when in fact it does not pass it, and a type II error would be committed. Table 12 shows the proportion of Type I and II errors, committed by the BIC test, for some parameter sets.

Table 12. Proportion of type I and II errors made by the BIC test.

α_2	Estimation Type I Error		Estimation Type II Error	
	$n = 32, m = 32$	$n = 64, m = 64$	$n = 32, m = 32$	$n = 64, m = 64$
0.05	0.031	0.043	0.077	0.155
0.01	0.008	0.009	0.022	0.044
0.001	0.001	0.001	0.002	0.011
0.0001	0	0	0	0

It can be seen that for $\alpha_2 = 0.0001$ type I and II errors are not made.

The outputs of RC4 [23] are known to pass numerous statistical tests [49], however they do not satisfy the BIC statistical test proposed in this work. This shows that the BIC statistical test complements the classic randomness tests, therefore it constitutes a tool to consider to evaluate stream ciphers.

5. Conclusions

An algorithm was proposed to extend the application of the Bit Independence Criterion (BIC) to stream ciphers. This algorithm detects the existence of statistical dependence between the inputs and outputs of a stream cipher. The effectiveness of the BIC test was experimentally confirmed when applied on random matrices, in which it does not detect dependence, and on the RC4 cipher, detecting statistical dependencies between the inputs and outputs of this cipher that are related with previously reported.

The algorithm depends on the number n of bits of the inputs, the number m of bits of the outputs, and the number l of inputs used. These parameters determine its complexity. The results achieved confirm the importance of varying the n and m parameters to apply the BIC criteria in the evaluation of stream ciphers. For RC4 the effectiveness of the criterion is a growing function of the n and m parameters.

It is recommended to guarantee the effectiveness of the proposed BIC test by selecting the values of the parameters greater than the minimum value estimated in the article. From that minimum, increase the values depending on the available computing power, estimating the time using the complexity expressions that were presented from the algorithm.

The BIC statistical test complements the classical statistical tests of randomness as it allows expanding the evaluation of the stream ciphers, by measuring the degree of independence present between the input of the cipher and its outputs, thus measuring other statistical characteristics that are not only the evaluation of randomness of their output sequences.

In future work, it is planned to apply this test to other stream ciphers, investigate the optimal choice of the m and n parameters and compare the effectiveness of the criterion taking into account the mutual information, the Pearson's coefficient, with the transformation mentioned, and the criteria applied in this work. The behavior of the proposal will be experimentally verified when the sample size increases. It will be investigated in an implementation variant using parallelism for Algorithm 3.

Author Contributions: Conceptualization, E.J.M.-C., G.S.-G. and C.M.L.-P.; methodology, E.J.M.-C., G.S.-G. and C.M.L.-P.; software, E.J.M.-C., G.S.-G. and C.M.L.-P.; validation, E.J.M.-C., R.S.-L. and O.R.; formal analysis, E.J.M.-C., G.S.-G., O.R., R.S.-L. and C.M.L.-P.; investigation, E.J.M.-C., G.S.-G., O.R., R.S.-L. and C.M.L.-P.; writing—original draft preparation, E.J.M.-C., G.S.-G., O.R., R.S.-L. and C.M.L.-P.; writing—review and editing, E.J.M.-C., G.S.-G., O.R., R.S.-L. and C.M.L.-P.; supervision, E.J.M.-C., G.S.-G., O.R., R.S.-L. and C.M.L.-P. All authors have read and agreed to the published version of the manuscript.

Funding: This research received no external funding.

Conflicts of Interest: The authors declare no conflict of interest.

Appendix A

Table A1. Notation table.

$f : \mathbb{F}_2^n \longrightarrow \mathbb{F}_2^m$	Describe the function that transforms n input bits into m output bits
$X = (x_1, \ldots, x_n)$	n-bit input binary vector
$Y = f(X) = (y_1, \ldots, y_m)$	m-bit output binary vector
e_i	Unit vector with 1 in the i-th component with $1 \leq i \leq n$
X^i	Vector resulting from the operation $X^i = X \oplus e_i$ for input X
Y^i	m-bit output binary vector corresponding to input X^i, $Y^i = f(X^i)$
$V^i = Y \oplus Y^i = (v_1^i, v_2^i, \ldots, v_m^i)$	Avalanche vector associated with vector e_i and input X
$v_j^i \in \mathbb{F}_2$	Avalanche variable associated to vector e_i and input X with $1 \leq j \leq m$
$D = \{X_1, \ldots, X_l\}$	Set of l inputs X_r, with $1 \leq r \leq l$
X_r^i	Vector resulting from the operation $X_r^i = X_r \oplus e_i$ for the input X_r

Table A1. *Cont.*

Y_r^i	Binary output vector of m bits corresponding to input X_r^i, $Y^i = f(X_r^i)$
$V_r^i = Y_r \oplus Y_r^i = (v_{r1}^i, v_{r2}^i, \ldots, v_{rm}^i)$	Avalanche vector associated with vector e_i and input X_r
$v_{rj}^i \in \mathbb{F}_2$	Avalanche variable associated to vector e_i and input X_r with $1 \leq j \leq m$

References

1. Marton, K.; Suciu, A.; Ignat, I. Randomness in digital cryptography: A survey. *Rom. J. Inf. Sci. Technol.* **2010**, *13*, 219–240.
2. Demirhan, H.; Bitirim, N. Statistical Testing of Cryptographic Randomness. *J. Stat. Stat. Actuar. Sci.* **2016**, *9*, 1–11.
3. ECRYPT Stream Cipher Project. 2011. Available online: http://cr.yp.to/streamciphers.html (accessed on 5 July 2020) [CrossRef]
4. Yerukala, N.; Kamakshi Prasad, V.; Apparao, A. Performance and statistical analysis of stream ciphers in GSM communications. *J. Commun. Softw. Syst.* **2020**, *16*, 11–18. [CrossRef]
5. Gorbenko, I.; Kuznetsov, A.; Lutsenko, M.; Ivanenko, D. The research of modern stream ciphers. In Proceedings of the 2017 4th International Scientific-Practical Conference Problems of Infocommunications. Science and Technology (PIC S&T), Kharkov, Ukraine, 10–13 October 2017; pp. 207–210. [CrossRef]
6. Upadhya, D.; Gandhi, S. Randomness evaluation of ZUC, SNOW and GRAIN stream ciphers. *Adv. Intell. Syst. Comput.* **2017**, *508*, 55–63. [CrossRef]
7. Rukhin, A.; Soto, J.; Nechvatal, J. *A Statistical Test Suite for Random and Pseudorandom Number Generators for Cryptographic Applications*; Technical Report April; Booz-Allen and Hamilton Inc.: Mclean, VA, USA, 2010.
8. Marsaglia, G. The Marsaglia Random Number CDROM Including the Diehard Battery of Tests of Randomness. Florida State University, 1995. Available online: http://stat.fsu.edu/pub/diehard/ (accessed on 5 July 2020).
9. L'ecuyer, P.; Simard, R. TestU01: A C library for empirical testing of random number generators. *ACM Trans. Math. Softw. TOMS* **2007**, *33*. [CrossRef]
10. McClellan, M.T.; Minker, J.; Knuth, D.E. *The Art of Computer Programming, Vol. 3: Sorting and Searching*; Addison-Wesley Professional: Boston, MA, USA, 1974; Volume 28, p. 1175. [CrossRef]
11. Shi, Z.; Zhang, B.; Feng, D.; Wu, W. Improved key recovery attacks on reduced-round Salsa20 and ChaCha. *Lect. Notes Comput. Sci.* **2013**, *7839 LNCS*, 337–351. [CrossRef]
12. Maitra, S.; Paul, G. New form of permutation bias and secret key leakage in keystream bytes of RC4. In *International Workshop on Fast Software Encryption*; Springer: Berlin/Heidelberg, Germany, 2008; Volume 5086 LNCS, pp. 253–269. [CrossRef]
13. Hancock, P.A. On the Design of Time. *Ergon. Des.* **2018**, *26*, 4–9. [CrossRef]
14. Qureshi, A.; Shah, T. S-box on subgroup of Galois field based on linear fractional transformation. *Electron. Lett.* **2017**, *53*, 604–606. [CrossRef]
15. Naseer, Y.; Shah, T.; Shah, D.; Hussain, S. A Novel Algorithm of Constructing Highly Nonlinear S-p-boxes. *Cryptography* **2019**, *3*, 6. [CrossRef]
16. Turan, M.S. On Statistical Analysis of Synchronous Stream Ciphers. *arXiv* **2008**, arXiv:1011.1669v3.
17. Duta, C.L.; Mocanu, B.C.; Vladescu, F.A.; Gheorghe, L. Randomness Evaluation Framework of Cryptographic Algorithms. *Int. J. Cryptogr. Inf. Secur.* **2014**, *4*, 31–49. [CrossRef]
18. Castro, J.C.H.; Sierra, J.M.; Seznec, A.; Izquierdo, A.; Ribagorda, A. The strict avalanche criterion randomness test. *Math. Comput. Simul.* **2005**, *68*, 1–7. [CrossRef]
19. Mishra, P.R.; Gupta, I.; Pillai, N.R. Generalized avalanche test for stream cipher analysis. In Proceedings of the International Conference on Security Aspects in Information Technology, Haldia, India, 19–22 October 2011; Volume 7011 LNCS, pp. 168–180. [CrossRef]
20. Srinivasan, C.; Lakshmy, K.V.; Sethumadhavan, M. Measuring diffusion in stream ciphers using statistical testing methods. *Def. Sci. J.* **2012**, *62*, 6–10. [CrossRef]
21. Sosa-Gómez, G.; Rojas, O.; Páez-Osuna, O. Using hadamard transform for cryptanalysis of pseudo-random generators in stream ciphers. *EAI Endorsed Trans. Energy Web* **2020**, *7*. [CrossRef]

22. Madarro Capó, E.J.; Cuellar, O.J.; Legón Pérez, C.M.; Gómez, G.S. Evaluation of input—Output statistical dependence PRNGs by SAC. In Proceedings of the 2016 International Conference on Software Process Improvement (CIMPS), Aguascalientes, Mexico, 12–14 October 2016; pp. 1–6. [CrossRef]
23. Paul, G.; Maitra, S. RC4: Stream cipher and its variants. *RC4 Stream Cipher Its Var.* **2011**, 1–281. [CrossRef]
24. Grosul, A.L.; Wallach, D.S. *A Related-Key Cryptanalysis of RC4*; Rice University: Houston, TX, USA, 2000; pp. 1–13.
25. Matsui, M. Key collisions of the RC4 stream cipher. In *International Workshop on Fast Software Encryption*; Springer: Berlin/Heidelberg, Germany, 2009; Volume 5665 LNCS, pp. 38–50. [CrossRef]
26. Chen, J.; Miyaji, A. How to find short RC4 colliding key pairs. In *International Conference on Information Security*; Springer: Berlin/Heidelberg, Germany, 2011; Volume 7001 LNCS, pp. 32–46. [CrossRef]
27. Maitra, S.; Paul, G.; Sarkar, S.; Lehmann, M.; Meier, W. New Results on Generalization of Roos-Type Biases and Related Keystreams of RC4. In *International Conference on Cryptology in Africa*; Springer: Berlin/Heidelberg, Germany, 2013; pp. 222–239. [CrossRef]
28. Maximov, A. *Some Words on Cryptanalysis of Stream Ciphers*; Citeseer: Lund, Sweden, 2006.
29. Vergili, I.; Yücel, M.D. Avalanche and bit independence properties for the ensembles of randomly chosen n × n s-boxes. *Turk. J. Electr. Eng. Comput. Sci.* **2001**, *9*, 137–145.
30. Karell-Albo, J.A.; Legón-Pérez, C.M.; Madarro-Capó, E.J.; Rojas, O.; Sosa-Gómez, G. Measuring independence between statistical randomness tests by mutual information. *Entropy* **2020**, *22*, 741. [CrossRef]
31. Ibrahim, H.; Khurshid, K. Performance Evaluation of Stream Ciphers for Efficient and Quick Security of Satellite Images. *Int. J. Signal Process. Syst.* **2019**, *7*, 96–102. [CrossRef]
32. Gorbenko, I.; Kuznetsov, A.; Gorbenko, Y.; Vdovenko, S.; Tymchenko, V.; Lutsenko, M. Studies on statistical analysis and performance evaluation for some stream ciphers. *Int. J. Comput.* **2019**, *18*, 82–88.
33. RC4 Cipher Is No Longer Supported in Internet Explorer 11 or Microsoft Edge. Available online: https://support.microsoft.com/en-us/help/3151631/rc4-cipher-is-no-longer-supported-in-internet-explorer-11-or-microsoft (accessed on 5 July 2020).
34. SSL Configuration Required to Secure Oracle HTTP Server after Applying Security Patch Updates. Available online: https://support.oracle.com/knowledge/Middleware/2314658_1.html (accessed on 5 July 2020).
35. Satapathy, A.; Livingston, J. A Comprehensive Survey on SSL/ TLS and Their Vulnerabilities. *Int. J. Comput. Appl.* **2016**, *153*, 31–38. [CrossRef]
36. Soundararajan, E.; Kumar, N.; Sivasankar, V.; Rajeswari, S. Performance analysis of security algorithms. In *Advances in Communication Systems and Networks*; Springer: Singapore, 2020; Volume 656, pp. 465–476. [CrossRef]
37. Jindal, P.; Makkar, S. Modified RC4 variants and their performance analysis. In *Microelectronics, Electromagnetics and Telecommunications*; Springer: Singapore, 2019; Volume 521, pp. 367–374. [CrossRef]
38. Parah, S.A.; Sheikh, J.A.; Akhoon, J.A.; Loan, N.A.; Bhat, G.M. Information hiding in edges: A high capacity information hiding technique using hybrid edge detection. *Multimed. Tools Appl.* **2018**, *77*, 185–207. [CrossRef]
39. Tyagi, M.; Manoria, M.; Mishra, B. Effective data storage security with efficient computing in cloud. *Commun. Comput. Inf. Sci.* **2019**, *839*, 153–164. [CrossRef]
40. Dhiman, A.; Gupta, V.; Singh, D. Secure portable storage drive: Secure information storage. *Commun. Comput. Inf. Sci.* **2019**, *839*, 308–316. [CrossRef]
41. Nita, S.; Mihailescu, M.; Pau, V. Security and Cryptographic Challenges for Authentication Based on Biometrics Data. *Cryptography* **2018**, *2*, 39. [CrossRef]
42. Zelenoritskaya, A.V.; Ivanov, M.A.; Salikov, E.A. Possible Modifications of RC4 Stream Cipher. *Mech. Mach. Sci.* **2020**, *80*, 335–341. [CrossRef]
43. Jindal, P.; Singh, B. Optimization of the Security-Performance Tradeoff in RC4 Encryption Algorithm. *Wirel. Pers. Commun.* **2017**, *92*, 1221–1250. [CrossRef]
44. Verdú, S. Empirical estimation of information measures: A literature guide. *Entropy* **2019**, *21*, 720. [CrossRef]
45. Hutson, A.D. A robust Pearson correlation test for a general point null using a surrogate bootstrap distribution. *PLoS ONE* **2019**, *14*. [CrossRef]
46. Liu, F.; Dong, Q.; Xiao, G. Probabilistic analysis methods of S-boxes and their applications. *Chin. J. Electron.* **2009**, *18*, 504–508.

47. Walpole, R.E.; Myers, R.H. *Probability & Statistics for Engineers & Scientists*; Pearson Education Limited: London, UK, 2012.
48. Siraj-Ud-Doulah, M. A Comparison among Twenty-Seven Normality Tests. *Res. Rev. J. Stat.* **2019**, *8*, 41–59.
49. Riad, A.M.; Shehat, A.R.; Hamdy, E.K.; Abou-Alsouad, M.H.; Ibrahim, T.R. Evaluation of the RC4 algorithm as a solution for converged networks. *J. Electr. Eng.* **2009**, *60*, 155–160.

Publisher's Note: MDPI stays neutral with regard to jurisdictional claims in published maps and institutional affiliations.

© 2020 by the authors. Licensee MDPI, Basel, Switzerland. This article is an open access article distributed under the terms and conditions of the Creative Commons Attribution (CC BY) license (http://creativecommons.org/licenses/by/4.0/).

Article

A Novel Intermittent Jumping Coupled Map Lattice Based on Multiple Chaotic Maps

Rong Huang [1,2], Fang Han [1,2,*], Xiaojuan Liao [3], Zhijie Wang [1,2] and Aihua Dong [1,2]

1. College of Information Science and Technology, Donghua University, Shanghai 201620, China; rong.huang@dhu.edu.cn (R.H.); wangzj@dhu.edu.cn (Z.W.); dongaihua@dhu.edu.cn (A.D.)
2. Engineering Research Center of Digitized Textile & Apparel Technology, Ministry of Education, Donghua University, Shanghai 201620, China
3. College of Information Science and Technology, Chengdu University of Technology, Chengdu 610059, China; liaoxiaojuan18@cdut.edu.cn
* Correspondence: yadiahan@163.com; Tel.: +86-021-67792315

Abstract: Coupled Map Lattice (CML) usually serves as a pseudo-random number generator for encrypting digital images. Based on our analysis, the existing CML-based systems still suffer from problems like limited parameter space and local chaotic behavior. In this paper, we propose a novel intermittent jumping CML system based on multiple chaotic maps. The intermittent jumping mechanism seeks to incorporate the multi-chaos, and to dynamically switch coupling states and coupling relations, varying with spatiotemporal indices. Extensive numerical simulations and comparative studies demonstrate that, compared with the existing CML-based systems, the proposed system has a larger parameter space, better chaotic behavior, and comparable computational complexity. These results highlight the potential of our proposal for deployment into an image cryptosystem.

Keywords: coupled map lattice; intermittent jumping; multi-chaos

1. Introduction

In 1985, Kaneko [1–5] formally proposed the Coupled Map Lattice (CML) consisting of discrete time steps, discrete spatial positions, and continuous state variables, and then deeply investigated its chaotic behavior. The series of studies [1–5] reveal that the CML system can create spatiotemporal chaos by combining local nonlinear dynamics and spatial diffusion. On the other hand, Fridrich [6] set forth that chaotic systems have remarkable cryptography properties, including simple structure, ergodicity, and high sensitivity to initial conditions and control parameters, and then constructed a chaos-based permutation-and-diffusion cipher architecture for encrypting images. These pioneering achievements have been continuous sources that inspire subsequent methods [7–29] and have enriched the field of chaotic cryptography.

Zhang and Wang [7,8] proposed a Non-adjacent Coupled Map Lattice (NCML) in which the spatial positions of coupled lattices are dynamically determined through a nonlinear Arnold cat map. Then, the spatiotemporal chaotic sequences were used to drive a bit-level group permutation phase. Guo et al. [9] redesigned the NCML system by replacing a Logistic map with a piecewise-linear chaotic map, in the hope of boosting the turbulence of local reaction. To increase the spatial diffusion, reference [10] provided a hybrid-coupling mechanism that considers the adjacent and non-adjacent interactions simultaneously. Later, Zhang et al. [11] employed a 3D Arnold cat map to establish coupling relations between lattices, which enlarges the chaotic regime over a wide range of parameters. In [12], Liu et al. developed a NCML-based S-box shuffling method, which strengthens the ability to resist linear password attacks. Wang et al. [13] put forward a Dynamically Coupled Map Lattice (DCML), in which the coupling coefficient dynamically varies with the spatiotemporal indices. Then, Wang et al. [14] combined the ideas behind NCML and DCML, and invented a Non-adjacent Dynamically Coupled Map Lattice (NDCML). In [15], the spatiotemporal chaotic sequences generated by the NDCML system were used to govern a bidirectional image diffusion phase. Along this line of thought, Tao et al. [16] constructed a tailor-made dynamical coupling architecture, and claimed that it can evenly spread diffusion energies over lattices. Reference [17] described a customized globally CML whose coupling term is a superposition of multiple chaotic maps. Zhang et al. introduced a fractional order Logistic map into the CML system, and investigated the spatiotemporal dynamics conditioned on non-adjacent coupling [18] or time delay [19], respectively. Inspired by [19], Lv et al. [20] defined a dedicated delay function, and developed a new CML system with time delay. This CML system [20] serves as a pseudo-random number generator for encrypting RGB images in a channel-crossed manner [21]. Wang et al. [22,23] focused on a kind of cross CML whose diffusion mechanism proceeds along the temporal and spatial dimensions simultaneously. A dynamical coupling coefficient controlled by Tent map [22] or a pair of module operations [23] was introduced into the cross CML in order to enhance the chaotic behaviors. Reference [24] presented a mixed CML, which organizes three chaotic maps together guided by the spatial positions of lattices. In addition, some works [25–29] extended the original CML system to two dimensions, and realized the deployment in some real-world applications like image encryption or hiding.

In summary, the ideas of the above works [7–11,13,14,16,17,22–24] mainly concentrate on three aspects: non-adjacent coupling [7–11], dynamical coupling [13,14,16,22,23], and multi-chaos mixing [17,24]. Based on our analysis, there still exist some problems in the three aspects. First, for the non-adjacent coupling, most of the previous works [7–11] merely adopted the Arnold cat map to determine the spatial positions of coupled lattices. However, we find that the Arnold cat map will repeatedly sample the same spatial position at a regular period. The repeated sampling usually tends to accumulate the diffusion energies heavily for a part of lattices, and thus incurs a non-uniform distribution of diffusion energies over lattices. Second, for the dynamical coupling, the previous works [13,14,16,22,23] simply endow the coupling coefficient with dynamics, but neglect the influence of coupling states on the spatiotemporal chaos. Third, for the multi-chaos mixing, the current works [17,24] fulfilled the mixing in a deterministic manner so that the mixing behavior can be clearly inferred according to the spatiotemporal indices. These problems not only limit the parameter space of spatiotemporal chaos, but also result in local chaotic behaviors, implying that some of the lattices may always lie in a non-chaotic regime. In such circumstances, extra computation or human labor may be required to carefully screen out the chaotic lattices before performing image encryption algorithms.

To alleviate the above problems, in this paper, we propose a novel intermittent jumping CML, abbreviated as IJCML, based on multiple chaotic maps. In our proposal, the pseudo-random information generated from multiple chaotic maps is integrated into the lattices in a complex nonlinear manner. The intermittent jumping mechanism is twofold. First, the intermittence refers to dynamically switching the coupling states between "off" and "on", which increases the complexity of diffusion. Second, the jumping establishes new non-adjacent coupling relations between lattices, which avoids the problem of repeated sampling so as to equalize the distribution of diffusion energies over lattices. We will describe the IJCML system at length in Section 2.

Our work studies the chaotic behaviors and nonlinear phenomena induced by the intermittent jumping mechanism, and observes the spatiotemporal chaos of the IJCML system from the following perspectives: diffusion energy analysis, power spectrum analysis, equilibrium degree analysis, information entropy analysis, inter-lattice independence analysis, Lyapunov exponent analysis, bifurcation diagram analysis, spatiotemporal behavior analysis, and computational complexity analysis. Note that these perspectives are widely chosen by [7–11,13,14,16,17,22–24] for analyzing a dynamical system. We will conduct extensive numerical simulations and comparative studies in Section 3. The simulation results demonstrate that the proposed IJCML system is superior to or comparable to the existing ones [1,7,13,14] in terms of the aforementioned analyses, and thus highlight the potential for deployment into a practical image cryptosystem.

2. Intermittent Jumping Coupled Map Lattice

There are three essential elements in the CML system [1]: coupling mechanism, nonlinear mapping function, and multiple processing units (lattices), which forms a reaction-diffusion architecture. In each time step, the reaction procedure updates the state variable of a lattice via the nonlinear mapping function. At the same time, the diffusion procedure establishes a coupling relation between the current lattice and a set of other lattices. The traditional CML system can be expressed as:

$$x_{n+1}(l) = (1-e) \cdot f(x_n(l)) + (e/2) \cdot [f(x_n(l-1)) + f(x_n(l+1))], \quad (1)$$

where $n = 1, 2, \cdots, N$ represent the time steps. The indices $l = 1, 2, \cdots, L$ stand for the spatial positions of lattices with a periodic boundary condition, and L is the size of the CML system. The symbol e denotes the coupling coefficient, whose value lies in the interval [0, 1]. In general, the Logistic map $x_{k+1} = f(x_k) = \mu \cdot x_k \cdot (1 - x_k)$ serves as the nonlinear mapping function, where the parameter $\mu \in (0, 4]$, and x_k represents the state variable after k iterations. Remarkably, when $\mu \in (3.57, 4]$, the Logistic map exhibits chaotic behaviors, including ergodicity and high sensitivity to initial conditions. See its bifurcation diagram and Lyapunov exponent in Figure 1a.

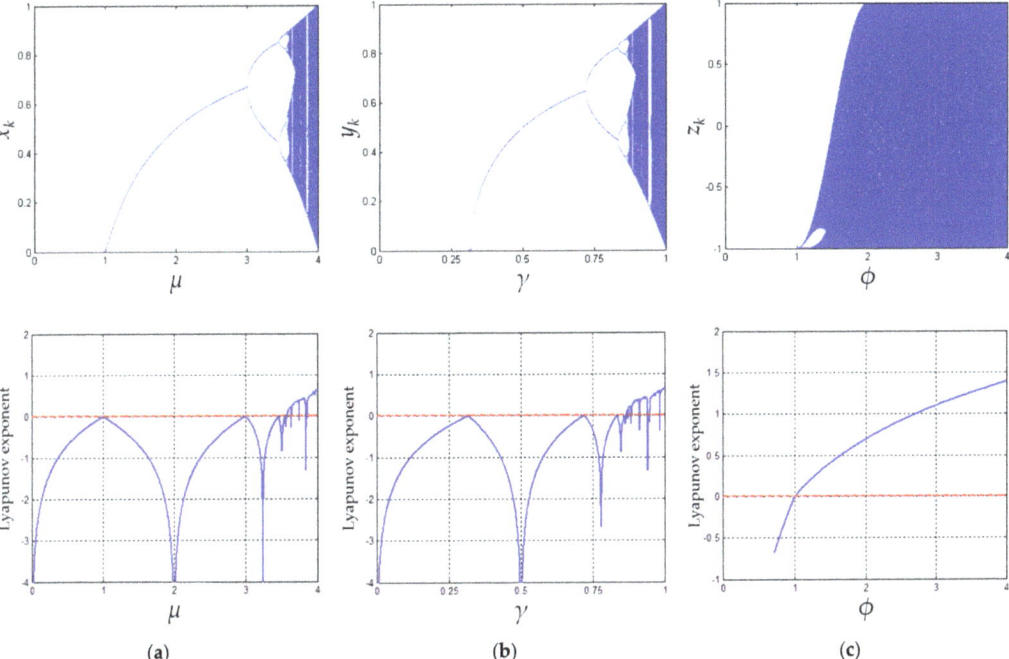

Figure 1. The bifurcation diagrams and Lyapunov exponents for (**a**) Logistic map, (**b**) Sine map, and (**c**) Chebyshev map, respectively.

Clearly, the coupling mechanism in Equation (1) is static, due to the constant coupling coefficient and the regularity of the coupling relations. To alleviate this problem, the NCML system [7] defines a non-adjacent coupling mechanism as follows:

$$x_{n+1}(l) = (1-e) \cdot f(x_n(l)) + (e/2) \cdot [f(x_n(u(l))) + f(x_n(v(l)))], \quad (2)$$

where the two spatial positions $u(l)$ and $v(l)$ are obtained from the Arnold cat map. That is:

$$\begin{bmatrix} u(l) \\ v(l) \end{bmatrix} = \begin{bmatrix} 1 & s \\ t & st+1 \end{bmatrix} \begin{bmatrix} l \\ l \end{bmatrix} \bmod(L) + \begin{bmatrix} 1 \\ 1 \end{bmatrix}, \quad (3)$$

where s and t are the parameters of the Arnold cat map. However, as discussed before, the Arnold cat map used in Equation (3) belongs to a time-invariant transformation, and tends to sample the same spatial position repeatedly within a single time step.

The DCML system [13] provides another way to break the static defect, which can be formulated as:

$$x_{n+1}(l) = (1 - \varepsilon_n(l)) \cdot f(x_n(l)) + (\varepsilon_n(l)/2) \cdot [f(x_n(l-1)) + f(x_n(l+1))], \quad (4)$$

where the coupling coefficient is no longer a constant, but varies with the spatiotemporal indices l and n. To be specific, the DCML system [13] leverages an auxiliary Logistic map of the form:

$$\varepsilon_{k+1} = f(\varepsilon_k) = \mu^{aux} \cdot \varepsilon_k \cdot (1 - \varepsilon_k), \quad (5)$$

where $k = 1, 2, \cdots, LN$, to generate the sequence of dynamical coupling coefficients. In practical use, this sequence is reshaped into a matrix of size $L \times N$ so as to gear towards the spatiotemporal indices in Equation (4). The auxiliary parameter μ^{aux} is set to 3.99 for achieving outstanding dynamics, and the initial value of ε_k is set to e.

The NDCML system [14], which combines the non-adjacent coupling mechanism and the dynamical coupling coefficient together, can be viewed as a natural extension of the NCML [7] and DCML [13] systems. That is:

$$x_{n+1}(l) = (1 - \varepsilon_n(l)) \cdot f(x_n(l)) + (\varepsilon_n(l)/2) \cdot [f(x_n(u(l))) + f(x_n(v(l)))], \quad (6)$$

where $u(l)$ and $v(l)$ are determined by Equation (3), and $\varepsilon_n(l)$ is obtained from the auxiliary Logistic map defined in Equation (5).

In this paper, we propose a novel Intermittent Jumping Coupled Map Lattice (IJCML), in which the pseudo-random information generated from the Logistic map, Sine map, and Chebyshev map will be integrated together in a dynamical manner. The definitions of the Sine map and the Chebyshev map can be described by the following equations:

$$y_{k+1} = g(y_k) = \gamma \cdot \sin(\pi \cdot y_k), \quad (7)$$

$$z_{k+1} = h(z_k) = \cos(\phi \cdot \arccos(z_k)), \quad (8)$$

where γ and ϕ are the systems' parameters. As shown in Figure 1b, the Sine map shares similar chaotic behaviors with the Logistic map. As shown in Figure 1c, z_k fills in the range $[-1, 1]$, and the Lyapunov exponent increases monotonically with ϕ. The Chebyshev map starts to enter the chaotic regime when $\phi \geq 1$.

With these preparations, we formulate the IJCML system as:

$$x_{n+1}(l) = \begin{cases} (1-e) \cdot f(x_n(l)) + (e/2) \cdot [g(x_n(u_n(l))) + g(x_n(v_n(l)))], & \text{if } w_n(l) \geq 0.5, \\ f(x_n(l)), & \text{otherwise,} \end{cases} \quad (9)$$

where $w_n(l)$ determines the couple state of the lth lattice at the nth time step. In this paper, the Chebyshev map is used to generate $w_n(l)$, taking the form $w_n(l) = (z_n(l) + 1)/2$, where $z_n(l)$ is the reshaped version of z_k. The parameter ϕ is set to 3.999 so that $w_n(l)$ varies with the spatiotemporal indices dynamically. The intermittence is reflected in the phenomenon that a lattice sometimes interacts with other ones (when $w_n(l) \geq 0.5$), and other times updates alone (when $w_n(l) < 0.5$). Clearly, the coupling states are mutually different between lattices.

When $w_n(l) \geq 0.5$, the spatial positions of coupled lattices are $u_n(l)$ and $v_n(l)$, respectively. In this paper, we abandon the Arnold cat map, and resort to a chaotic map to determine $u_n(l)$ and $v_n(l)$. Specifically, the pseudo-random information $w_n(l)$ is reused here in the following form:

$$\begin{cases} u_n(l) = mod(\lfloor w_n(l) \cdot 10^7 \rfloor, L) + 1, \\ v_n(l) = mod(\leftrightharpoons \lfloor w_n(l) \cdot 10^7 \rfloor, L) + 1, \end{cases} \quad (10)$$

where the floor sign $\lfloor \cdot \rfloor$ rounds down to the nearest integer of the number enclosed within the sign. The sign $\leftrightharpoons$ taking its right operand, namely $\lfloor w_n(l) \cdot 10^7 \rfloor$ in Equation (10), as input represents a compound S-box substitution operation. Specifically, the right operand is first converted to a six-digit hexadecimal integer. For example, if $\lfloor w_n(l) \cdot 10^7 \rfloor = 1234567$, its hexadecimal representation equals '12D687'. Second, the six-digit hexadecimal integer is divided into three groups, each of which contains two digits. The three groups of '12D687' are '12', 'D6', and '87', respectively. Third, apply substitution operation to each group in turn, where we use the S-box recommended in AES [30]. See Figure 2 for details. By doing so, we obtain 'C9', 'F6', and '17', respectively. Forth, concatenate the substitution values, yielding 'C9F617'. Convert this six-digit hexadecimal integer into decimal representation, yielding 13235735. As shown in Equation (10), the compound S-box substitution operation is only used in the course of calculating $v_n(l)$, and the nonlinearity of the S-box ensures that $u_n(l) \neq v_n(l)$ holds. This is a straightforward way to avoid sampling the same spatial position for a specific spatiotemporal index. The pseudo-randomness of $u_n(l)$ and $v_n(l)$ inherits from $w_n(l)$, so that they vary with the spatiotemporal indices as well. The new non-adjacent coupling relations are time-varying, compared with the one in NCML [13] or NDCML [14]. In this paper, we use the term "jumping" to characterize the pseudo-randomness and the spatiotemporal variability of the coupling relation.

		0	1	2	3	4	5	6	7	8	9	A	B	C	D	E	F
	0	63	7C	77	7B	F2	6B	6F	C5	30	01	67	2B	FE	D7	AB	76
	1	CA	82	C9	7D	FA	59	47	F0	AD	D4	A2	AF	9C	A4	72	C0
	2	B7	FD	93	26	36	3F	F7	CC	34	A5	E5	F1	71	D8	31	15
	3	04	C7	23	C3	18	96	05	9A	07	12	80	E2	EB	27	B2	75
	4	09	83	2C	1A	1B	6E	5A	A0	52	3B	D6	B3	29	E3	2F	84
	5	53	D1	00	ED	20	FC	B1	5B	6A	CB	BE	39	4A	4C	58	CF
	6	D0	EF	AA	FB	43	4D	33	85	45	F9	02	7F	50	3C	9F	A8
x	7	51	A3	40	8F	92	9D	38	F5	BC	B6	DA	21	10	FF	F3	D2
	8	CD	0C	13	EC	5F	97	44	17	C4	A7	7E	3D	64	5D	19	73
	9	60	81	4F	DC	22	2A	90	88	46	EE	B8	14	DE	5E	0B	DB
	A	E0	32	3A	0A	49	06	24	5C	C2	D3	AC	62	91	95	E4	79
	B	E7	C8	37	6D	8D	D5	4E	A9	6C	56	F4	EA	65	7A	AE	08
	C	BA	78	25	2E	1C	A6	B4	C6	E8	DD	74	1F	4B	BD	8B	8A
	D	70	3E	B5	66	48	03	F6	0E	61	35	57	B9	86	C1	1D	9E
	E	E1	F8	98	11	69	D9	8E	94	9B	1E	87	E9	CE	55	28	DF
	F	8C	A1	89	0D	BF	E6	42	68	41	99	2D	0F	B0	54	BB	16

Figure 2. S-box of AES. The substitution value is determined by the intersection of the row with index 'x' and the column with index 'y'.

As shown in Equations (9) and (10), the proposed IJCML system combines the Logistic map, the Sine map, and the Chebyshev map together in a nonlinear complex way. First, the Logistic map serves as the nonlinear mapping function as usual. Second, the Sine map plays a key role in the coupling term. In this paper, the parameter of the Sine map is set to 0.999, so that, in each time step, the state variables of the lattices at the spatial positions $u_n(l)$ and $v_n(l)$ will be updated along chaotic trajectories. Third, the Chebyshev map generates the pseudo-random signal that is used to ensure the dynamic switching of coupling states and coupling relations. Additionally, as long as we protect the initial state variables of the Sine map and the Chebyshev map (e.g., treating them as keys), it is virtually impossible for an adversary to infer the mixing behaviors. This alleviates the problem of deterministic mixing in the existing works [17,24]. This work provides an intermittent jumping mechanism, which can effectively promote the turbulence evolution amongst lattices.

3. Analysis of Dynamic Behaviors

In this section, we analyze the chaotic behaviors of the IJCML system from nine aspects, and make comparisons with the baseline systems, including CML [1], NCML [7], DCML [13], and NDCML [14]. Throughout our analysis, the number of lattices L is set to 100, while the maximum number of iterations, namely N, equals 1000. Unless explicitly stated, the auxiliary parameters, including s, t, μ^{aux}, are set to 23, 12, 3.99, respectively.

3.1. Diffusion Energy Analysis

The goal of this analysis is to check whether the distribution of diffusion energies between lattices is uniform or not. In this paper, the diffusion energy of the lth lattice is defined as:

$$E(l) = \sum_{l'=1}^{L} \sum_{n'=1}^{N} \frac{\delta(l - u_{n'}(l')) + \delta(l - v_{n'}(l'))}{2} \left[\varepsilon_{n'}(l')\right]^2, \quad (11)$$

where $\delta(\cdot)$ denotes the Dirac delta function. Equation (11) reveals that there are two factors determining the diffusion energy. One is the total number of times that the lth lattice is sampled after N iterations. Another is the square of the coupling coefficient. Note that, for a specific CML-based system, the spatiotemporal variability of $u_{n'}(l')$, $v_{n'}(l')$, and $\varepsilon_{n'}(l')$ in Equation (11) may be absent. For example, for the original CML system, we shall set that $u_{n'}(l') = l' - 1$, $v_{n'}(l') = l' + 1$, and $\varepsilon_{n'}(l') = e$, respectively. Since this analysis is primarily concerned with whether the distribution is uniform or not, rather than the magnitude of $E(l)$, we normalize the diffusion energy as follows:

$$\overline{E}(l) = E(l) / \max_{l} E(l), \quad (12)$$

where $l = 1, 2, \cdots, L$. It is desirable to equalize this distribution so that each lattice contributes equally to the diffusion of spatiotemporal dynamics.

As demonstrated in [7,8], the non-adjacent coupling mechanism can strengthen the diffusion of spatiotemporal dynamics to some extent. Hence, in this paper, we define inter-lattice diffusion distances as follows:

$$D(l) = \frac{\sum_{r=\{u,v\}} \sum_{l'=1}^{L} \sum_{n'=1}^{N} \delta(l - r_{n'}(l')) \cdot \min(|l - r_{n'}(l')|, L - |l - r_{n'}(l')|)}{\sum_{r=\{u,v\}} \sum_{l'=1}^{L} \sum_{n'=1}^{N} \delta(l - r_{n'}(l'))}, \quad (13)$$

which measures how far the lth lattice's diffusion energy is spread to the others. Note that, for the IJCML system, the calculations of Equations (11) and (13) are triggered only when $w_{n'}(l') \geq 0.5$. It is desirable for larger inter-lattice diffusion distances with uniform distribution.

Figure 3 exhibits results for the diffusion energy analysis. In this simulation study, we set e to 0.99. The blue histogram in the upper panel is the normalized distribution of diffusion energies, while the bottom panel shows the histogram of the inter-lattice diffusion distances. In addition, we print the standard deviation (std^-) of the blue histogram, and print the average (avg^+) and the standard deviation (std^-) of the orange one, where the superscript "+" (or "−") is intended to indicate that a higher (or lower) value is better.

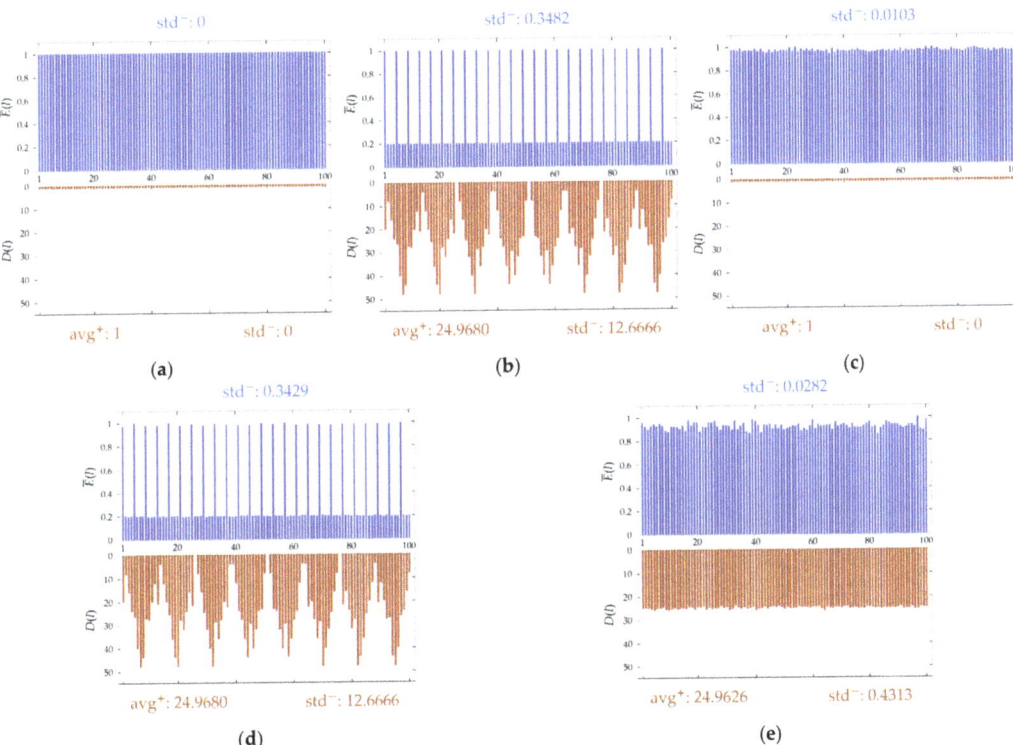

Figure 3. Diffusion energy analysis for (**a**) the CML system, (**b**) the NCML system, (**c**) the DCML system, (**d**) the NDCML system, and (**e**) the proposed IJCML system, respectively.

We find that although the CML system achieves the best uniformity with zero-valued standard deviations, $D(l)$ equals 1 for all l, reflecting that each lattice only interacts with its adjacent counterpart in a regular way. As shown in Figure 3b, the inter-lattice diffusion distances reach larger values, which demonstrates that the NCML system indeed enlarges the range of coupling. However, there are several peaks that appear periodically in the histograms. This is due to the use of the Arnold cat map, which repeatedly samples the same spatial position at a regular period. Such non-uniform distribution implies that only several lattices may dominate the diffusion of spatiotemporal dynamics, leading to the local chaotic behaviors. As shown in Figure 3c, the dynamical coupling coefficient can slightly modulate the diffusion energies. However, it fails to equalize NDCML's histograms. In contrast, the proposed IJCML system equalizes the distribution of diffusion energies, and achieves larger inter-lattice diffusion distances at the same time. This validates that the IJCML system can uniformly spread the diffusion energies to non-adjacent lattices, so as to encourage the diffusion of spatiotemporal dynamics.

3.2. Power Spectrum Analysis

Power spectrum is a significant tool to visualize whether a sequence has chaotic characteristics or not. The Fourier theorem sets forth that any periodic signal can be expressed as Fourier series, corresponding to a discrete frequency spectrum. By contrast, an aperiodic signal is represented by Fourier integral with a continuous frequency spectrum. As such, if the power spectrum has one or more distinct peaks, the signal must be periodic or quasi-periodic. Conversely, if the power spectrum has a noise-like modality without obvious peaks, the signal is chaotic.

Figure 4 shows results for the power spectrum analysis, in which we investigate twelve combinations of settings for the parameters μ and e. In this simulation study, the sequence with 1000 elements generated by 50th lattice is selected for analysis. In addition, for the sake of observation, we swap the left and right halves of a power spectrum, in the sense of shifting the zero-frequency component to the center.

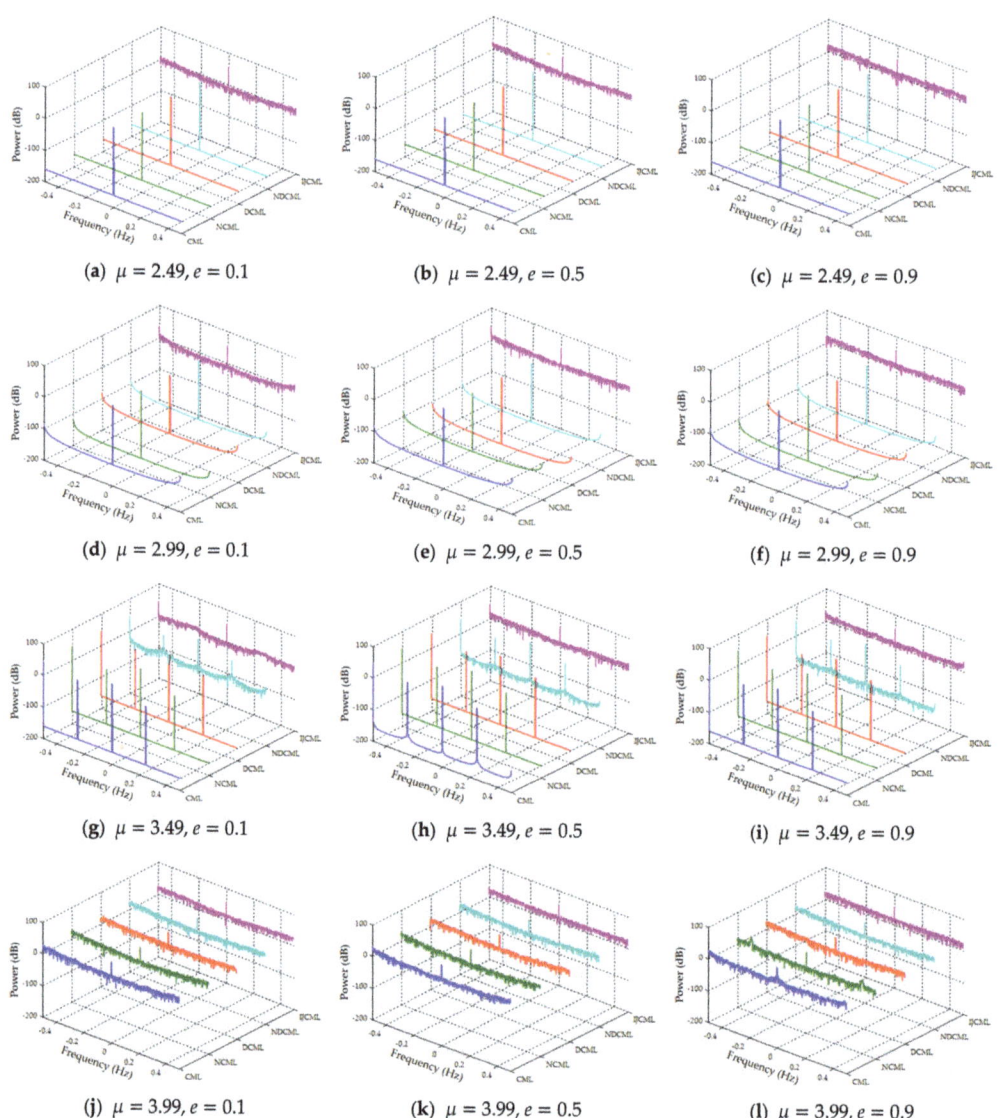

Figure 4. Power spectrum analysis for the CML system, the NCML system, the DCML system, the NDCML system, and the proposed IJCML system.

By comparison, we find that the parameter e has a relatively small influence on the power spectrum. Thus, the following discussion mainly focuses on the effect of different values of μ. As we see, for the baseline systems, almost all the power is concentrated at zero-frequency component when $\mu = 2.49$ or 2.99. This means that there is almost no fluctuation in the corresponding sequence, so that it belongs to a direct-current signal. When $\mu = 3.49$, the baseline systems distribute the power onto several frequency components. This implies that the corresponding sequence contains periodic transitions between several state variables, so that it belongs to a deterministic signal. When $\mu = 3.99$, five power spectrums share the similar noise-like modality, meaning that all the systems have turned into the complete turbulence pattern [4]. Remarkably, even when $\mu = 2.49$, 2.99, or 3.49, the IJCML system's power is evenly dispersed over all frequency components, in the sense that there exists only one small peak submerged in the noise-like fluctuations. This phenomenon demonstrates that the proposed system behaves better than the baseline ones, and indeed enlarges the parameter space of spatiotemporal chaos.

3.3. Equilibrium Degree Analysis

As the name implies, the equilibrium degree analysis is to evaluate whether the number of ones is close to the number of zeros in a binary sequence. Naturally, its definition can be written in the following form:

$$ED(l) = 1 - |P(l) - Q(l)|/N, \qquad (14)$$

where $l = 1, 2, \cdots, L$. The notations $P(l)$ and $Q(l)$ record the number of ones and zeros, respectively, in the lth binary sequence, and we have $N = P(l) + Q(l)$. The value of equilibrium degree lies in the interval $[0, 1]$, and the higher the value, the better the equilibrium degree. For some bit-level image cryptosystems, the equilibrium degree analysis is helpful to check whether a CML-based system can generate pseudo-random binary sequences.

Figure 5 shows results for the equilibrium degree analysis under the setting $2.8 \leq \mu < 4$ and $0 \leq e \leq 1$. In this simulation study, the sequence $x_n(l)$ generated by the lth lattice is first binarized through the operation $mod(\lfloor x_n(l) \cdot 10^7 \rfloor, 2)$, where $n = 1, 2, \cdots, N$. We calculate the average value of $ED(l)$ over all lattices, where $l = 1, 2, \cdots, L$.

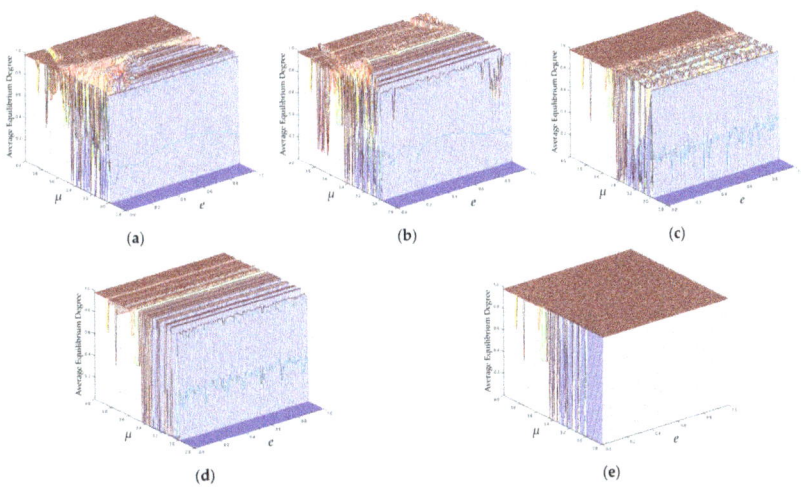

Figure 5. Equilibrium degree analysis for (**a**) the CML system, (**b**) the NCML system, (**c**) the DCML system, (**d**) the NDCML system, and (**e**) the proposed IJCML system, respectively.

As we see, when $\mu < 2.99$, the average values of equilibrium degrees are all equal to 0 for the baseline systems. When $2.99 \leq \mu \leq 3.57$, the baseline systems become trapped in an unstable transition zone, over which the average values are drastically varying. When $\mu > 3.57$, the DCML system reaches a flat plateau with the average value close to 1 as shown in Figure 5c. For the CML system, however, there exists an obvious sunken area embedded into the flat plateau at the range $3.57 < \mu < 3.99$ and $0 < e < 0.18$. For the NDCML system, two straight ravines occur at $\mu = 3.74$ and 3.84, respectively, and directly go through the flat plateau. The NCML system has a wavelike plateau, which involves multiple sunken areas and ravines. This means that these baseline systems fail to achieve a stable equilibrium degree even when the nonlinear mapping function $f(x)$ has entered into the chaotic regime (i.e., $\mu > 3.57$). In contrast, due to the mixture of multiple chaotic maps, the proposed IJCML system possesses a more stable and larger flat plateau as shown in Figure 5e. If e is set to 0, the IJCML system's equilibrium degree degrades considerably, which validates that the intermittent jumping mechanism is helpful to enhance the chaotic behaviors. Based on these observations, we state that the IJCML system outperforms the baseline ones in terms of the equilibrium degree.

3.4. Information Entropy Analysis

Information entropy is a frequently-used criterion to measure the degree of confusion of a dynamical system. Following the quantization procedure previously applied in [13,14,16], we uniformly quantize a state variable $x_n(l)$ to an integer between 0 and 9. Specifically, the quantization procedure takes the form $\lfloor x_n(l) \cdot 10 \rfloor$. The information entropy of a specific lattice is defined by:

$$H(\mathbf{c}) = - \sum_{c=0}^{9} p(c) log_2 p(c), \tag{15}$$

where $c = \{c | c = 0, 1, \cdots, 9\}$ denotes the possible levels of quantization, and $p(c)$ represents the probability of occurrence of c in a quantized sequence. Since we consider ten levels in the course of quantizing, the maximum value of $H(\mathbf{c})$ equals $log_2 10 \approx 3.3219$. It is desirable to achieve a higher value of $H(\mathbf{c})$, which implies better pseudo-randomness of the quantized sequence.

Figure 6 shows the results for the information entropy analysis. We conduct the simulation study under the setting $3 \leq \mu < 4$ and $0 \leq e \leq 1$, and calculate the average value of $H(\mathbf{c})$ over all L lattices.

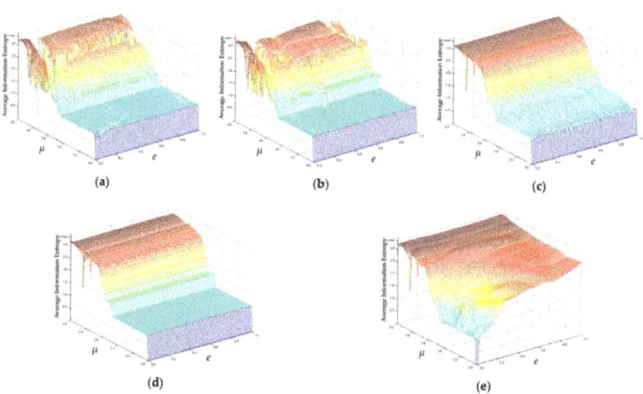

Figure 6. Information entropy analysis for (**a**) the CML system, (**b**) the NCML system, (**c**) the DCML system, (**d**) the NDCML system, and (**e**) the proposed IJCML system, respectively. The symbol 'max' on the vertical axis signifies the maximum value of the average information entropy, namely 3.3219.

We see that, for the baseline systems, the average values of $H(\mathbf{c})$ form a stair-stepped upward trend as μ increases, and the two notable steps appear at $\mu = 3.02$ and 3.48, respectively. When $\mu > 3.48$, the DCML system has a smooth surface, in which the average values are independent of e. For the NDCML system, two straight ravines that appear at $\mu = 3.74$ and 3.84, respectively, go through the surface. This phenomenon is consistent with the observation in Section 3.3, demonstrating that there still exist defects in the NDCML system even when μ is set to a higher value. For the CML system, a L-shaped valley with larger area is embedded into the surface. This analysis provides a warning that one should carefully select the parameters before using a CML-based image cryptosystem. Equipped with the non-adjacent coupling mechanism, the NCML system reduces the area of the valley, but fails to construct a smooth surface. For the IJCML system, the average value of $H(\mathbf{c})$ increases along the axes of μ and e, and the leaf-like surface covers a larger parameter space of spatiotemporal chaos. These results verify that the IJCML system is superior to the baseline ones in terms of information entropy.

3.5. Inter-Lattice Independence Analysis

Some CML-based image cryptosystems may use multiple lattices simultaneously for driving the permutation and diffusion phases. In such circumstance, the higher the independence between two lattices, the stronger the ability to combat information leakage. In this paper, a Mutual-Information-based Entropy Distance (MIED) is defined to measure the independence between two quantized sequences. The mathematical formula is:

$$MIED(\mathbf{c}^i, \mathbf{c}^j) = MAX - I(\mathbf{c}^i; \mathbf{c}^j), \tag{16}$$

where the superscript i (or j) stands for the lattice's index, while the symbol MAX is a constant, representing the maximum possible value of $H(\mathbf{c}^i)$ (or $H(\mathbf{c}^j)$). In this simulation study, we quantize a state variable into ten levels as before, so that $MAX = log_2 10 \approx 3.3219$. In Equation (16), $I(\mathbf{c}^i; \mathbf{c}^j)$ is the mutual information between $\mathbf{c}^i$ and $\mathbf{c}^j$, and its definition is written as:

$$I(\mathbf{c}^i; \mathbf{c}^j) = \sum_{c^i=0}^{9} \sum_{c^j=0}^{9} p(c^i, c^j) log_2 \frac{p(c^i, c^j)}{p(c^i) \cdot p(c^j)}. \tag{17}$$

The nonnegativity of mutual information ensures that $0 \leq MIED(\mathbf{c}^i, \mathbf{c}^j) \leq MAX$ holds for any pair of lattices i and j. The upper and lower bounds, namely MAX and 0, correspond to the maximal and minimal degrees of independence, respectively.

Figure 7 shows the resulting values of MIED for the inter-lattice independence analysis, in which four combinations of settings for μ and e, namely {3.49, 0.1}, {3.49, 0.9}, {3.99, 0.1} and {3.99, 0.9}, are considered.

When $\mu = 3.49$, the values of MIED, in general, form a middle-level plane containing fluctuations of various patterns. Interestingly, for the NCML system, the punctate leakages are expanded to box-like pits as e increases. This is because a greater coupling coefficient usually enhances the correlations between a pair of lattices i and j, thereby yielding a higher value of $I(\mathbf{c}^i; \mathbf{c}^j)$. Moreover, as shown in the second column of Figure 7b, the box-like pits are arranged with a regular period. This arises from the fact that the Arnold cat map, which is used in the NCML system, forces some pairs of lattices to interact with each other repeatedly during the iterative procedure. Unfortunately, this defect is inherited by the NDCML system, as shown in Figure 7d.

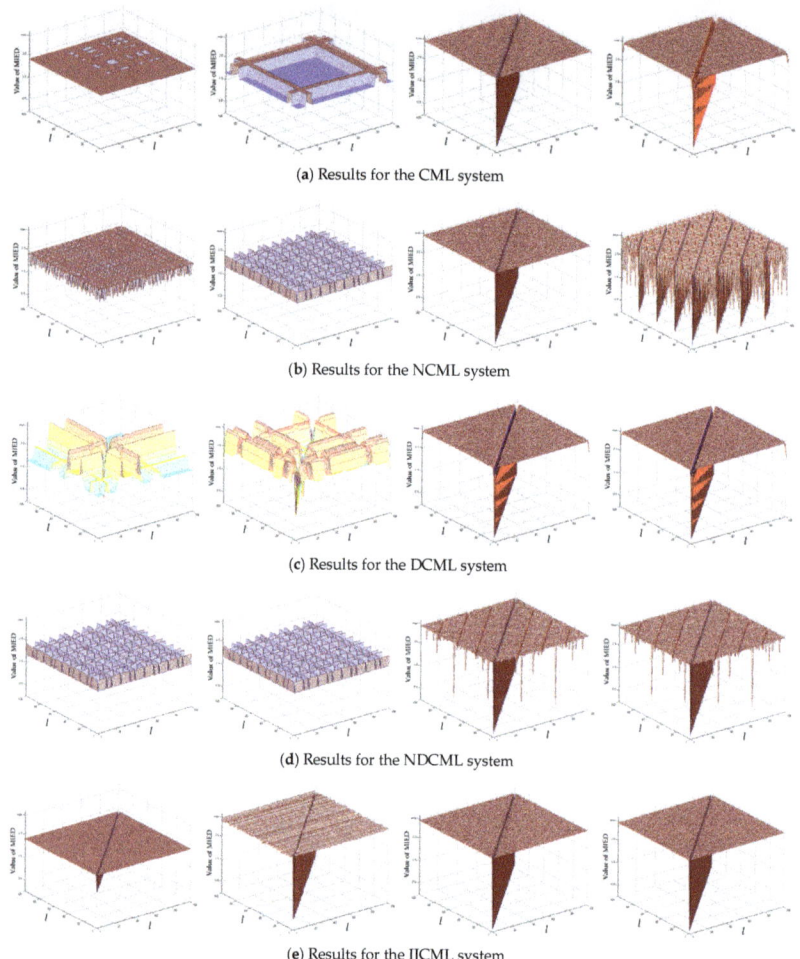

Figure 7. Resulting values of MIED for the inter-lattice independence analysis. The four settings of $\{\mu, e\}$, from left to right, are $\{3.49, 0.1\}$, $\{3.49, 0.9\}$, $\{3.99, 0.1\}$, and $\{3.99, 0.9\}$, respectively. The symbol 'max' on the vertical axis signifies the maximum value of MIED, namely 3.3219.

When $\mu = 3.99$, the values of MIED constitute a plateau lying close to *MAX*. The groove along the principal diagonal corresponds to the self-independence of a specific lattice, namely $\text{MIED}(\mathbf{c}^i, \mathbf{c}^i)$, where $i = 1, 2, \cdots, L$. Clearly, a flat plateau with a narrow groove indicates good performance of the inter-lattice independence. For the CML system, increasing e reinforces the interactions between two lattices, and thus results in a wider groove and four sunken regions around the corners. A similar phenomenon also happens in the DCML system, implying that a dynamical coupling coefficient is insufficient to eliminate the coupling-caused correlations. For the NCML and NDCML systems, there exist multiple diagonal grooves embedded into the flat plateau due to the use of Arnold cat map. From the first two columns of Figure 7e, we find that the IJCML system can reach higher value of MIED when $\mu = 3.49$ and $e = 0.9$. More significantly, the IJCML system forms the flattest plateau with a unit-width groove when $\mu = 3.99$. These comparisons demonstrate that the IJCML system has a larger parameter space for generating the independent sequences randomly.

Further, we calculate the average values of MIED over all lattices via Equation (18) as follows:

$$\frac{\sum_{i=1}^{L} \sum_{j=1}^{L,\ i \neq j} MIED(\mathbf{c}^i, \mathbf{c}^j)}{L \cdot (L-1)}, \tag{18}$$

and investigate the inter-lattice independence under the setting $3.5 \leq \mu < 4$ and $0 \leq e \leq 1$. The corresponding results are displayed in Figure 8.

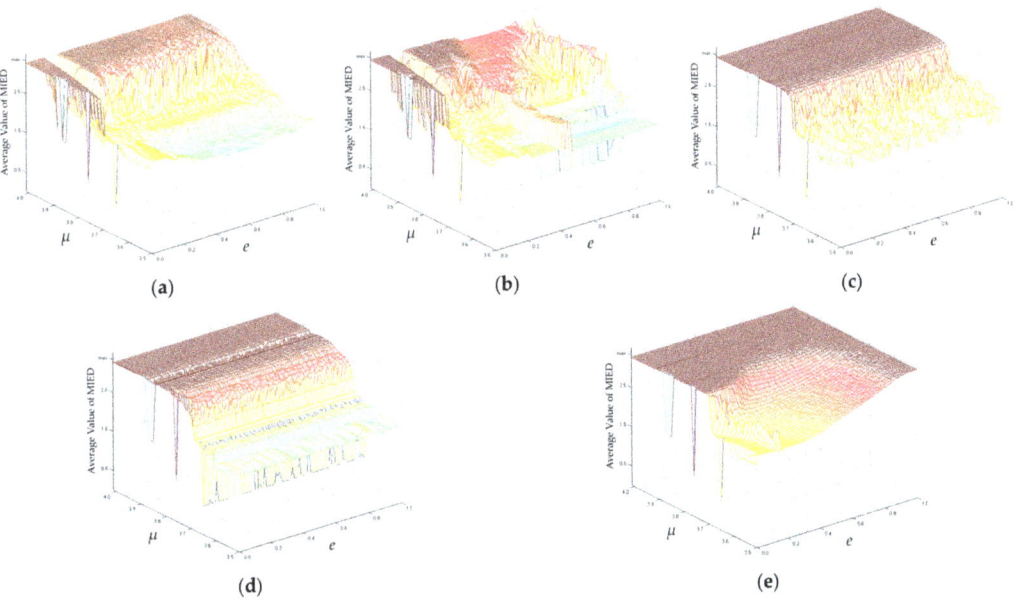

Figure 8. Average values of MIED for the inter-lattice independence analysis. (**a**) Result for the CML system. (**b**) Result for the NCML system. (**c**) Result for the DCML system. (**d**) Result for the NDCML system. (**e**) Result for the proposed IJCML system. The symbol 'max' on the vertical axis signifies the maximum value of the average information entropy, namely 3.3219.

As we see, the CML and NCML systems share the same defect, in the sense that a cuboid-shaped valley appears at the range $0.12 < e < 0.18$. This result suggests that the CML and NCML systems may be unsuitable for a multi-lattice-based image cryptosystem. It is clear that the DCML and NDCML systems are well-behaved when $\mu > 3.7$, and are both insensitive to e. In most cases, the IJCML system achieves higher values than the baseline systems, especially when $3.5 < \mu < 3.7$ and $0.85 < e \leq 1$. This result reflects that the intermittent jumping mechanism can effectively compensate for the performance degeneration, and thus demonstrates that the IJCML system surpasses the baseline ones in terms of the inter-lattice independence.

3.6. Lyapunov Exponent Analysis

Lyapunov exponent [31] is a canonical metric in describing a dynamical system. It quantifies the average exponential rate of separation (or convergence) between two initially close trajectories along each direction of phase space. Mathematically, the definition of Lyapunov exponent is formulated as:

$$\lambda = \lim_{n \to \infty} \frac{1}{n} \sum_{i=1}^{n} \ln \left| \frac{dF(x)}{dx} \right|_{x=x_i}, \tag{19}$$

where λ denotes the Lyapunov exponent of $F(x)$, while i stands for the time step. Typically, a positive Lyapunov exponent indicates that $F(x)$ has chaotic behaviors with the trajectories diverging exponentially.

Kolmogorov-Sinai Entropy Density (KSED), which incorporates the Lyapunov exponents of all lattices, is devoted to measuring the overall dynamics of a CML-based system. Formally, its definition is written as:

$$KSED = \sum_{l=1}^{L} \lambda^+(l)/L, \qquad (20)$$

where $\lambda^+(l)$ represents the positive Lyapunov exponent of the lth lattice. In other words, Equation (20) overlooks all negative Lyapunov exponents during its calculation.

In addition, Kolmogorov-Sinai Entropy Breadth (KSEB) [32] is used, in this paper, to count the proportion of chaotic lattices. We calculate its value as follows:

$$KSEB = L^+/L, \qquad (21)$$

where L^+ denotes the number of lattices whose Lyapunov exponents are positive. Clearly, the higher the values of KSED and KSEB, the better chaotic behaviors of a dynamical system.

Figure 9 exhibits the resulting values of KSED under the setting $3 \leq \mu < 4$ and $0 \leq e \leq 1$. Starting from $\mu = 3.57$, the DCML system possesses a smooth surface with positive curvatures, validating that increasing μ can constantly strengthen the chaotic behaviors. For the NDCML system, however, there exist two objectionable ravines appearing at $\mu = 3.74$ and 3.84, and straightly passing through the surface. The CML and NCML systems behave unstably when $\mu > 3.57$, as the values of KSED are varying sharply and a distinct valley appears at the range $3.57 < \mu < 3.99$ and $0.11 < e < 0.16$. The IJCML system achieves numerous high values of KSED that form a flat plateau. Noticeably, even when $\mu < 3.57$, setting e to a higher value can still lead the system into the chaotic regime. Thus, the proposed system has a larger parameter space of spatiotemporal chaos, which benefits from the intermittent jumping mechanism.

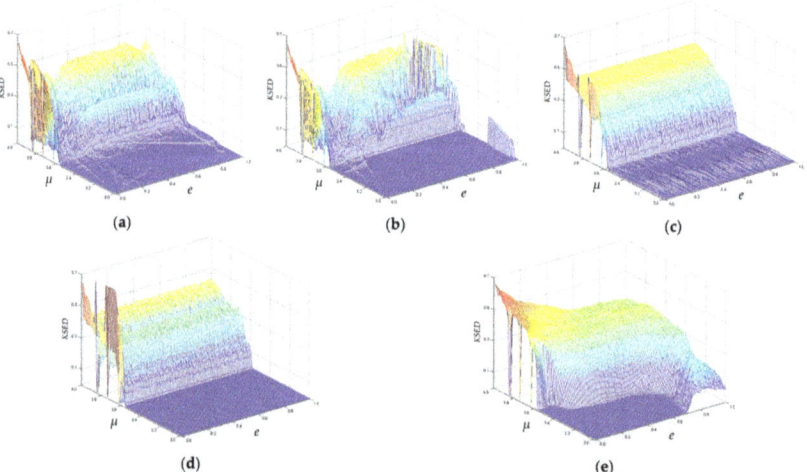

Figure 9. Resulting values of KSED for the Lyapunov exponent analysis. (**a**) Result for the CML system. (**b**) Result for the NCML system. (**c**) Result for the DCML system. (**d**) Result for the NDCML system. (**e**) Result for the proposed IJCML system.

In Figure 10, we display the resulting values of KSEB. The maximum value of KSEB, being equal to 1.0, means that all the lattices in the system are activated into spatiotemporal chaos. At first glance, the IJCML system's flat plateau, as shown in Figure 10e, is of the largest area. We note that, among the baseline systems, only the DCML system has a stable flat plateau when $\mu > 3.57$. For the other baseline systems, there exist deep valleys or straight ravines embedded into the flat plateau. In other words, even though μ is set to a higher value, some lattices are still far from spatiotemporal chaos. Therefore, one shall spend time inspecting the dynamics carefully for each lattice before deploying them in an image cryptosystem.

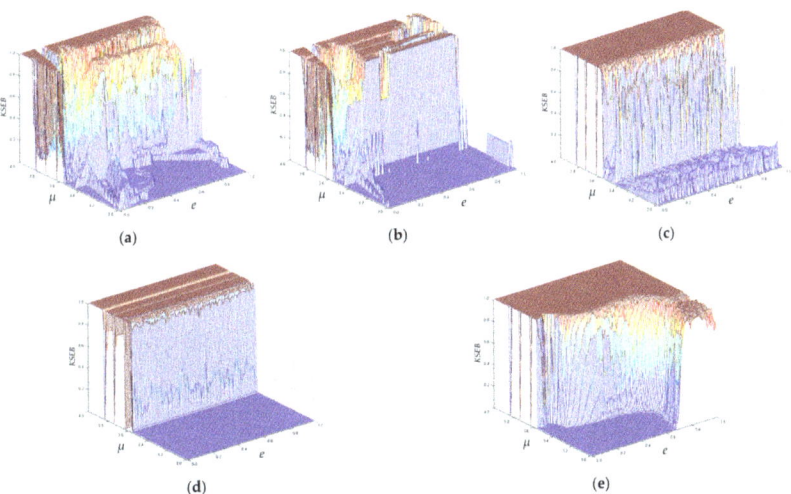

Figure 10. Resulting values of KSEB for the Lyapunov exponent analysis. (**a**) Result for the CML system. (**b**) Result for the NCML system. (**c**) Result for the DCML system. (**d**) Result for the NDCML system. (**e**) Result for the proposed IJCML system.

Further, we design a statistical tool called Cumulative Percentage Function (CPF) for analyzing $KSEB$. The CPF is to reflect the percentage of parameter pairs $\{\mu, e\}$ making $KSEB$ less than or equal to a specific threshold α. We show the plots of CPFs in Figure 11. The ideally optimal case is that all possible combinations of μ and e can lead a system into the fully chaotic regime, where all its lattices have the chaotic behaviors, namely $KSEB = 1.0$. Obviously, the ideally optimal case corresponds to an impulse function centered at 1.0. Among all CPFs, the pink one is closest to the impulse function, consisting of a wide flat region and a narrow step region. The higher the step occurs at $\alpha = 1.0$ in Figure 11, the larger the area of flat plateau in Figure 10. Based on above observations, we state that the proposed IJCML system is better than the baseline ones in terms of the Lyapunov exponent analysis.

3.7. Bifurcation Diagram Analysis

The bifurcation diagram, which is a plot of the state variables versus the control parameter(s), is commonly used to analyze a dynamical system. Let all the systems, in this simulation study, work under the setting $3.4 \leq \mu < 4$ and $e = \{0.1, 0.5, 0.9\}$. In Figure 12, we depict the bifurcation diagrams, in which the state variables are harvested from the 1st, 50th, and 100th lattices. For convenient comparison, the bifurcation diagrams belonging to different systems are coated with different colors.

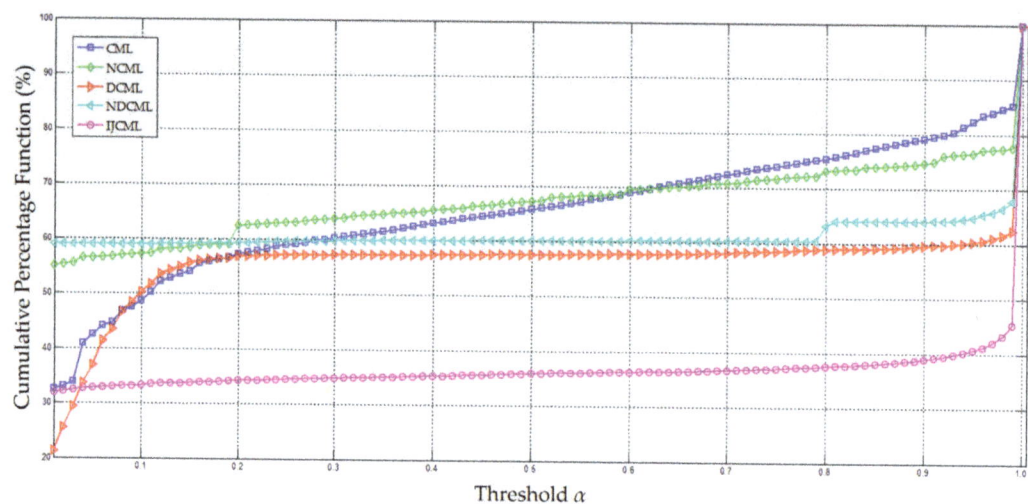

Figure 11. Plots of the cumulative percentage functions.

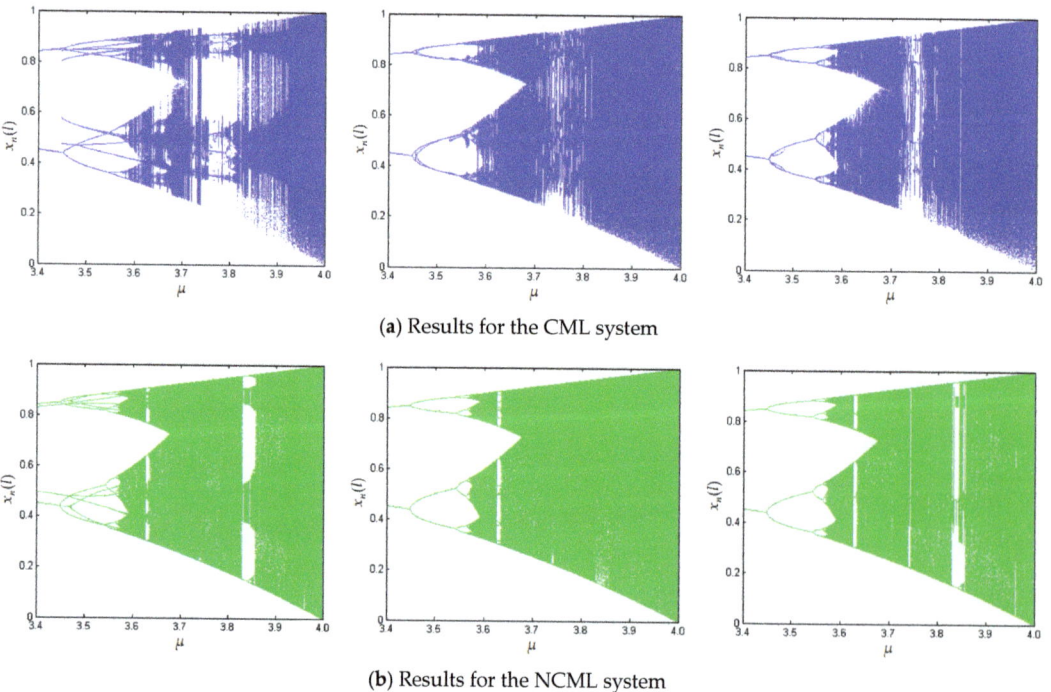

(**a**) Results for the CML system

(**b**) Results for the NCML system

Figure 12. *Cont.*

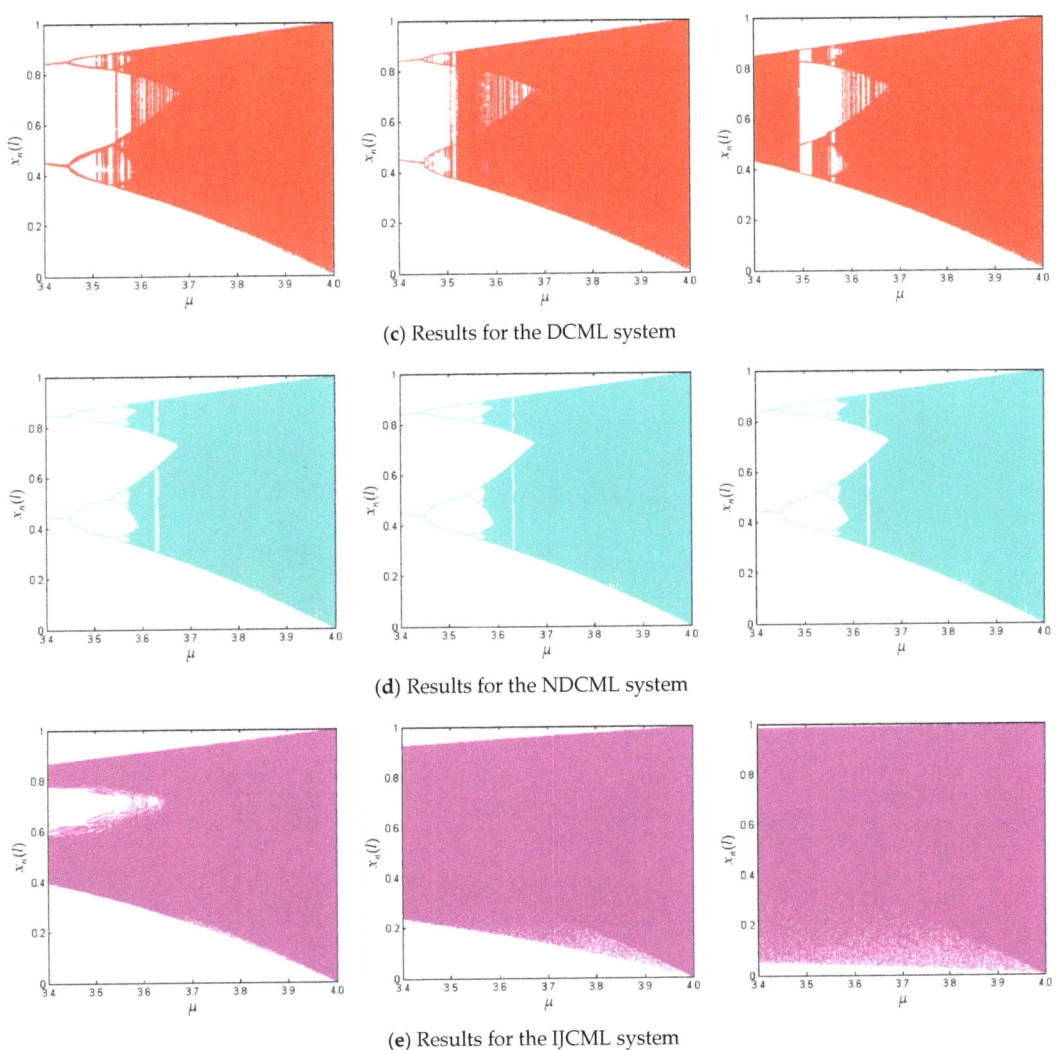

Figure 12. Bifurcation diagrams. The setting of e, from left to right, is 0.1, 0.5, and 0.9, respectively.

For the baseline systems, we can clearly observe that period-doubling bifurcations form a route to spatiotemporal chaos. For example, the NDCML system gets trapped in period-two oscillations when $\mu = 3.4$. Two bifurcation points at $\mu = 3.45$ double the periods of the orbits and give rise to pitchfork bifurcations. This process is repeated as μ increases. Starting from $\mu = 3.57$, it is virtually impossible to observe the pitchfork bifurcations because the number of bifurcation points becomes large and the gaps between the bifurcation points are negligible. Unfortunately, the chaotic oscillations of the baseline systems may occasionally be interspersed with periodic windows in some cases. For example, when setting μ to 3.63, we can observe the periodic windows in the NDCML system, where the oscillations suddenly degenerate into several periods. Comparing the baseline systems' bifurcation diagrams, we find that they roughly share the same phenomenon. That is, the period-doubling bifurcations trigger chaotic oscillations interspersed with periodic windows. By contract, the proposed IJCML system has a large number of bifurcation points with tiny gaps between them even when $\mu = 3.4$. This demonstrates that the IJCML system enters the chaotic regime early. Additionally, the IJCML system greatly reduces the periodic windows compared with the baseline ones. As shown in Figure 12e, a higher coupling coefficient introduces more pseudo-random motions from the multiple chaotic maps, so as to further enhance the chaotic oscillations. Based on these comparisons, we claim that the IJCML system behaves better than the baseline ones in terms of the bifurcation diagram analysis.

3.8. Spatiotemporal Behavior Analysis

Spatiotemporal behavior is a means of testing diffusions between lattices. In this simulation study, e is fixed at 0.5 for simplicity, while μ is set to 3.15, 3.57, and 3.99, respectively, for adjusting the spatiotemporal behaviors.

In Figure 13, we show space-amplitude plots, which are used to analyze the spatiotemporal behaviors. Similarly, the space-amplitude plots belonging to different systems are coated with different colors.

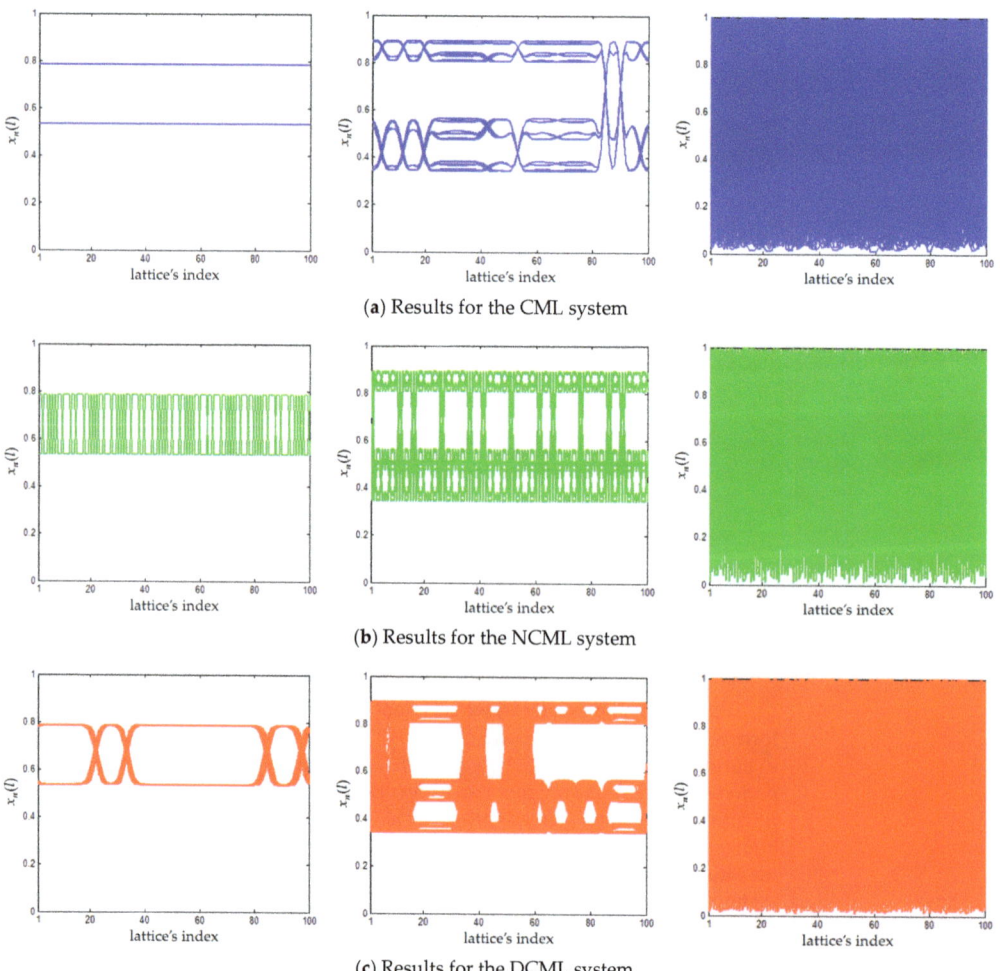

(a) Results for the CML system

(b) Results for the NCML system

(c) Results for the DCML system

Figure 13. Cont.

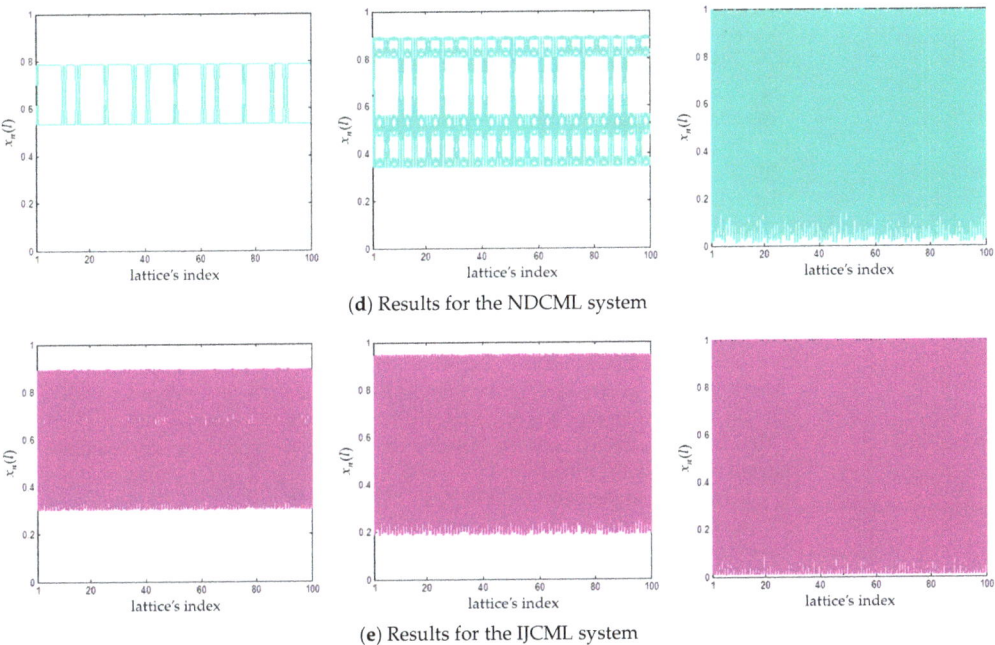

(d) Results for the NDCML system

(e) Results for the IJCML system

Figure 13. Space-amplitude plots for spatiotemporal behavior analysis. The setting of μ, from left to right, is 3.15, 3.57, and 3.99, respectively.

When $\mu = 3.15$, the CML system simply creates period-two responses. The other baseline systems twist the period-two responses at a regular pace, resulting in a X-shaped pattern. The twisting points correspond to stable solutions of a system, whose positions depend on the initial state variables. In contrast, the X-shaped pattern is unrecognizable for the IJCML system. These results reveal that, when the baseline systems are restricted to the frozen random pattern [4], the IJCML system has advanced into the defect chaotic diffusion pattern [4].

When $\mu = 3.57$, we can observe the period-doubling behaviors from the space-amplitude plots of the baseline systems, in the sense that the long-thin X-shaped pattern evolves into a compound version. Specifically, there exist multi-period responses twisted alternately, and the dynamic range of the solutions has been enlarged at the same time. In particular, some of the lattices in the DCML system, for example, the 40th one, have entered the defect chaotic diffusion pattern [4]. These results reveal that the baseline systems are transitioning from the frozen random pattern [4] to the defect chaotic diffusion pattern [4].

When $\mu = 3.00$, the period-changing behaviors become complex and unstable, resulting in extremely twisted and superimposed responses. Remarkably, the dynamic range is further extended towards the upper and lower bounds. These results reveal that all the systems have entered the complete turbulence pattern [4], while the IJCML system has the strongest ability to approach the lower bound. Consequently, we state that the IJCML system has better spatiotemporal behaviors than the baseline ones.

3.9. Computational Complexity Analysis

Computational complexity reflects the time and space consumptions while executing a target algorithm so as to provide a guidance for developing practical software.

First, we theoretically analyze the time complexity. Like the baseline systems [1,7,13,14], the proposed IJCML system also has two nested "for" loops in its calculation procedure. The outer "for" loop runs the dynamical system by N iterations, while the inner one traverses the L lattices. Hence, all the five CML-based systems investigated in this paper have the quadratic time complexity, namely $\mathcal{O}(NL)$.

Second, we theoretically analyze the space complexity. Clearly, it is necessary to save the resulting pseudo-random matrix of size $N \times L$. This requires quadratic space complexity, namely $\mathcal{O}(NL)$. Compared with the original CML system [1], the variants, i.e., the NCML system [7], the DCML system [13], the NDCML system [14], and the proposed IJCML system need to occupy additional space to save the intermediate variables obtained from the auxiliary chaotic maps. See Equations (3), (5), and (8) for details. Fortunately, since these auxiliary chaotic maps can be separately executed out of the two nested "for" loops, the final space complexity remains the same, namely $\mathcal{O}(NL)$.

However, it is undeniable that the variants may require more time and space consumptions in practice. To make an intuitive comparison, we experimentally count the average time for running 1000 iterations over 100 lattices, and inspect the size of the additional space. The parameters μ and e are set to $[3, 3.99]$ and $[0, 0.99]$, respectively, and both of them take 0.01 as step size. Our computing device is a desktop computer with a 2.90 GHz Intel i7-10700 central processing unit, 16.00 GB memory. Our programming environment is Matlab R2017a installed on the Window 10 operation system. The intermediate variables are saved as ".mat" format.

Table 1 lists the simulation results. The IJCML system takes 0.5400 s, on average, to generate a pseudo-random matrix of size 1000×100, which is faster than the NCML system [7], the DCML system [13], and the NDCML system [14], and is comparable to the original CML system [1]. High efficiency of the IJCML system is due to the intermittent jumping mechanism because the coupling operation, as shown in Equation (9), will be skipped with a probability of about 50%. On the other hand, the NCML system [7] only requires 3.42 KB space to additionally save the time-invariant spatial positions. In contrast, the remaining three variants prepare about 746.00 KB space to accommodate the dynamical coupling coefficients [13,14] or the intermediate variables $w_n(l)$ obtained from the Chebyshev map. Fortunately, a space of size 746.00 KB is almost negligible for modern hardware devices. Consequently, the computational complexity of the proposed IJCML is better than or comparable to the baseline systems [1,7,13,14].

Table 1. Time consumption and size of the additional space.

Items	CML [1]	NCML [7]	DCML [13]	NDCML [14]	IJCML (Ours)
Time consumption	0.4891 s	0.6160 s	0.6027 s	0.6239 s	0.5400 s
Size of the additional space	none	3.42 KB	746.00 KB	749.45 KB	744.00 KB

4. Conclusions

In this paper, we propose a novel IJCML system based on multiple chaotic maps. The intermittent jumping mechanism establishes a new coupling mode, in which not only the coupling states but also the coupling relations dynamically vary with the spatiotemporal indices. We conduct extensive numerical simulations and comparative studies to analyze the proposed system from the following aspects: the diffusion energy analysis, the power spectrum analysis, the equilibrium degree analysis, the information entropy analysis, the inter-lattice independence analysis, the Lyapunov exponent analysis, the bifurcation diagram analysis, the spatiotemporal behavior analysis, and the computational complexity analysis. The simulation results adequately demonstrate that, compared with the baseline systems [1,7,13,14], the proposed IJCML system has better chaotic behaviors, which brings stronger spatiotemporal chaos for a single lattice and higher independency between lattices. Our future work is to study the effective way for discretizing the IJCML system, and to consider the issues of practical realization and circuit implementation [33].

Author Contributions: Conceptualization, R.H. and F.H.; methodology, R.H., F.H., and Z.W.; project administration, R.H., F.H., and X.L.; funding acquisition, R.H., F.H., and X.L.; software, R.H. and A.D.; writing, R.H.; validation, X.L. and A.D. All authors have read and agreed to the published version of the manuscript.

Funding: This research was funded in part by the Fundamental Research Funds for the Central Universities (17D110408), the National Natural Science Foundation of China (11972115, 62001099, 61806171), and the National Key Research and Development Program of China (2019YFC1521300).

Institutional Review Board Statement: Not applicable.

Informed Consent Statement: Not applicable.

Data Availability Statement: Data sharing not applicable.

Conflicts of Interest: The authors declare no conflict of interest.

References

1. Kaneko, K. Spatiotemporal intermittency in coupled map lattices. *Prog. Theor. Phys.* **1985**, *74*, 1033–1044. [CrossRef]
2. Kaneko, K. Turbulence in coupled map lattices. *Phys. D* **1986**, *18*, 475–476. [CrossRef]
3. Kaneko, K. Lyapunov analysis and information flow in couple map lattices. *Phys. D* **1986**, *23*, 436–447. [CrossRef]
4. Kaneko, K. Pattern dynamics in spatiotemporal chaos: Pattern selection, diffusion of defect and pattern competition intermittency. *Phys. D* **1989**, *34*, 1–41. [CrossRef]
5. Kaneko, K. Overview of coupled map lattices. *Chaos* **1992**, *2*, 279–282. [CrossRef] [PubMed]
6. Fridrich, J. Symmetric ciphers based on two-dimensional chaotic maps. *Int. J. Bifurcat. Chaos* **1998**, *8*, 1259–1284. [CrossRef]
7. Zhang, Y.Q.; Wang, X.Y. A symmetric image encryption algorithm based on mixed linear-nonlinear coupled map lattice. *Inf. Sci.* **2014**, *273*, 329–351. [CrossRef]
8. Zhang, Y.Q.; Wang, X.Y. A new image encryption algorithm based on non-adjacent coupled map lattices. *Appl. Soft Comput.* **2015**, *26*, 10–20. [CrossRef]
9. Guo, S.F.; Liu, Y.; Gong, L.H.; Yu, W.Q.; Gong, Y.L. Bit-level image cryptosystem combining 2D hyper-chaos with a modified non-adjacent spatiotemporal chaos. *Multimed. Tools Appl.* **2018**, *77*, 21109–21130. [CrossRef]
10. Wang, X.Y.; Zhao, H.Y.; Wang, M.X. A new image encryption algorithm with nonlinear-diffusion based on multiple coupled map lattices. *Opt. Laser Technol.* **2019**, *115*, 42–57. [CrossRef]
11. Zhang, H.; Wang, X.Q.; Xie, H.W.; Wang, C.P.; Wang, X.Y. An efficient and secure image encryption algorithm based on non-adjacent couple maps. *IEEE Access* **2020**, *8*, 122104–122120. [CrossRef]
12. Liu, L.Y.; Zhang, Y.Q.; Wang, X.Y. A novel method for constructing the S-box based on spatiotemporal chaotic dynamics. *Appl. Sci.* **2018**, *8*, 2650. [CrossRef]
13. Wang, X.Y.; Feng, L.; Wang, S.B.; Zhang, C.; Zhang, Y.Q. Spatiotemporal chaos in coupled logistic map lattice with dynamic coupling coefficient and its application in image encryption. *IEEE Access* **2018**, *6*, 39705–39724.
14. Wang, X.Y.; Feng, L.; Li, R.; Zhang, F.C. A fast image encryption algorithm based on non-adjacent dynamically coupled map lattice model. *Nonlinear Dyn.* **2019**, *95*, 2797–2824. [CrossRef]
15. Wang, X.Y.; Zhao, H.Y.; Feng, L.; Ye, X.L.; Zhang, H. High-sensitivity image encryption algorithm with random diffusion based on dynamic-coupled map lattices. *Opt. Laser Eng.* **2019**, *122*, 225–238. [CrossRef]
16. Tao, Y.; Cui, W.H.; Zhang, Z. Spatiotemporal chaos in multiple dynamically coupled map lattices and its application in a novel image encryption algorithm. *J. Inf. Secur. Appl.* **2020**, *55*, 102650. [CrossRef]
17. Wang, X.Y.; Qin, X.M.; Liu, C.M. Color image encryption algorithm based on customized globally coupled map lattices. *Multimed. Tools Appl.* **2019**, *78*, 6191–6209. [CrossRef]
18. Zhang, Y.Q.; Wang, X.Y.; Liu, L.Y.; He, Y.; Liu, J. Spatiotemporal chaos of fractional order logistic equation in nonlinear coupled lattices. *Commun. Nonlinear Sci. Numer. Simulat.* **2017**, *52*, 52–61. [CrossRef]
19. Zhang, Y.Q.; Wang, X.Y.; Liu, L.Y.; Liu, J. Fractional order spatiotemporal chaos with delay in spatial nonlinear coupling. *Int. J. Bifurcat. Chaos* **2018**, *28*, 1850020. [CrossRef]
20. Lv, X.P.; Liao, X.F.; Yang, B. A novel pseudo-random number generator from coupled map lattice with time-varying delay. *Nonlinear Dyn.* **2018**, *94*, 325–341. [CrossRef]
21. Lv, X.P.; Liao, X.F.; Yang, B. Bit-level plane image encryption based on coupled map lattice with time-varying delay. *Mod. Phys. Lett. B* **2018**, *32*, 1850124. [CrossRef]
22. Wang, M.X.; Wang, X.Y.; Wang, C.P.; Xia, Z.Q.; Zhao, H.Y.; Gao, S.; Zhou, S.; Yao, N.M. Spatiotemporal chaos in cross coupled map lattice with dynamic coupling coefficient and its application in bit-level color image encryption. *Chaos Soliton. Fract.* **2020**, *139*, 110028. [CrossRef]
23. Wang, M.X.; Wang, X.Y.; Zhao, T.T.; Zhang, C.; Xia, Z.Q.; Yao, N.M. Spatiotemporal chaos in improved cross coupled map lattice and its application in a bit-level image encryption scheme. *Inf. Sci.* **2021**, *544*, 1–24. [CrossRef]

24. Wang, X.Y.; Guan, N.N.; Zhao, H.Y.; Wang, S.W.; Zhang, Y.Q. A new image encryption scheme based on coupling map lattices with mixed multi-chaos. *Sci. Rep.* **2020**, *10*, 9784. [CrossRef]
25. Zhang, Y.Q.; He, Y.; Wang, X.Y. Spatiotemporal chaos in mixed linear-nonlinear two-dimensional coupled logistic map lattice. *Physica A* **2018**, *490*, 148–160. [CrossRef]
26. Lu, G.Q.; Smidtaite, R.; Howard, D.; Ragulskis, M. An image hiding scheme in a 2-dimensional coupled map lattice of matrices. *Chaos Soliton Fract.* **2019**, *124*, 78–85. [CrossRef]
27. Kumar, S.; Kumar, R.; Kumar, S.; Kumar, S. Cryptographic construction using coupled map lattice as a diffusion model to enhanced security. *J. Inf. Secur. Appl.* **2019**, *46*, 70–83. [CrossRef]
28. He, Y.; Zhang, Y.Q.; Wang, X.Y. A new image encryption algorithm based on two-dimensional spatiotemporal chaotic system. *Neural Comput. Appl.* **2020**, *32*, 247–260. [CrossRef]
29. Zou, C.Y.; Wang, X.Y.; Li, H.F.; Wang, Y.Z. Enhancing the kinetic complexity of 2-D digital coupled chaotic lattice. *Nonlinear Dyn.* **2020**, *102*, 2925–2943. [CrossRef]
30. FIPS PUB 197. *Advanced Encryption Standard*; National Institute of Standards and Technology: Gaithersburg, MD, USA, 2001.
31. Wolf, A.; Swift, J.B.; Swinney, H.L.; Vastano, J.A. Determining Lyapunov exponents from a time-series. *Phys. D* **1985**, *16*, 285–317. [CrossRef]
32. Zhang, Y.Q.; Wang, X.Y.; Liu, J.; Chi, Z.L. An image encryption scheme based on the MLNCML system using DNA sequences. *Opt. Laser Eng.* **2016**, *82*, 95–103. [CrossRef]
33. Kocarev, L. Chaos-based cryptography: A brief overview. *IEEE Circ. Syst. Mag.* **2001**, *1*, 6–21. [CrossRef]

 *applied sciences*

Article

A Digital Cash Paradigm with Valued and No-Valued e-Coins

Ricard Borges [1,2,†] and Francesc Sebé [1,2,*,†]

1 Department of Mathematics, Universitat de Lleida, E-25001 Lleida, Spain; rborges@matematica.udl.cat
2 Cybercat: Center for Cybersecurity Research of Catalonia, E-25001 Lleida, Spain
* Correspondence: francesc.sebe@udl.cat; Tel.: +34-973-702-713
† These authors contributed equally to this work.

Abstract: Digital cash is a form of money that is stored digitally. Its main advantage when compared to traditional credit or debit cards is the possibility of carrying out anonymous transactions. Diverse digital cash paradigms have been proposed during the last decades, providing different approaches to avoid the double-spending fraud, or features like divisibility or transferability. This paper presents a new digital cash paradigm that includes the so-called no-valued e-coins, which are e-coins that can be generated free of charge by customers. A vendor receiving a payment cannot distinguish whether the received e-coin is valued or not, but the customer will receive the requested digital item only in the former case. A straightforward application of bogus transactions involving no-valued e-coins is the masking of consumption patterns. This new paradigm has also proven its validity in the scope of privacy-preserving pay-by-phone parking systems, and we believe it can become a very versatile building block in the design of privacy-preserving protocols in other areas of research. This paper provides a formal description of the new paradigm, including the features required for each of its components together with a formal analysis of its security.

Keywords: cryptography; digital cash; privacy

Citation: Borges, R.; Sebé, F. A Digital Cash Paradigm with Valued and No-Valued e-Coins. *Appl. Sci.* **2021**, *11*, 9892. https://doi.org/10.3390/app11219892

Academic Editors: Safwan El Assad, René Lozi and William Puech

Received: 22 September 2021
Accepted: 20 October 2021
Published: 22 October 2021

Publisher's Note: MDPI stays neutral with regard to jurisdictional claims in published maps and institutional affiliations.

Copyright: © 2021 by the authors. Licensee MDPI, Basel, Switzerland. This article is an open access article distributed under the terms and conditions of the Creative Commons Attribution (CC BY) license (https://creativecommons.org/licenses/by/4.0/).

1. Introduction

The European Commission defines digital cash (also referred to as *e-money* or *e-cash*) as a digital alternative to cash. It allows users to make cashless payments with money stored on a card or a phone, or over the Internet.

Digital cash was first proposed in the early 1980s by Chaum [1] in a proposal based on the newly invented blind signature cryptographic primitive. That proposal includes three actors: the *bank*, the *payer* (the *customer*), and the *payee* (the *vendor*), and three protocols: *withdraw*, *spend*, and *deposit*. An e-coin is a random sequence of data which has been digitally signed by the bank. During the withdraw protocol, the payer generates a random sequence and, after paying for it, asks the bank to digitally sign it through a blind signature protocol. This ensures that the bank does not learn any information about the issued e-coin so that the payer will be able to spend it anonymously in the future. An e-coin is spent by means of a process by which the payer transmits it to the payee. Finally, when the payee deposits an e-coin, they receive its monetary value from the bank.

Chaum's proposal [1] is both *anonymous* (the identity of the payer is not revealed during payment) and *unlinkable* (it is not possible to determine whether two payments were made by the same payer or not).

A transaction using a digital currency occurs entirely digitally. This means that a dishonest payer could store a copy of an e-coin after having spent it and then try to spend it again in the future. This is the *double-spending* fraud. Two main strategies have been proposed to cope with attempts to double-spend. In [1], upon receiving a payment, the payee asks the bank to check that the received e-coin has not been spent before. This is an *on-line* double-spending fraud prevention strategy. In the *off-line* alternative [2], double spending is not checked during the payment process, but it can be detected later when the e-coin is deposited. In cases of double-spending, the anonymity of the double-spender is lifted so that they can be prosecuted.

Research on digital cash systems has addressed diverse features such as e-coin *divisibility*, which allows a user to withdraw a coin and spend it several times by dividing its value (up to a given limit). The first practical divisible digital cash system was proposed in [3]. That proposal is anonymous but it is possible to link several spends from a single divisible coin. The system in [4] is divisible and unlinkable but the vendor and the bank know which part of the coin is being spent. Hence, unlinkability is provided partially. The first divisible and strong unlinkable e-cash system was described in [5]. A more efficient system in terms of computational and communications cost was proposed later in [6]. Research in the divisibility feature is still active with proposals such as [7], in which a large e-coin can be divided into several small ones with arbitrarily integer values. The system is efficient as both the pay and deposit procedures run in constant time. The authors of [8] provide a formal and complete security model for divisible e-cash and study constructions based on pseudo-random functions.

Another research line focused on digital cash systems able to deal with low value payments, that is, *micropayments* [9]. These systems have been addressed by the research community with proposals aiming to reduce the number of public key operations by replacing them with lightweight hash computations. The Micromint system [9] can be seen as a precursor of the widely-known *proof-of-work* concept in modern cryptocurrencies. Although digital cash systems need to be efficient both in terms of computation and communication costs, the ever increasing capacity of computers and mobile devices has eliminated the need for so many lightweight proposals. For instance, the authors of [10] achieve a reduced cost by using elliptic curve public key cryptography.

Transferability is the most difficult feature of paper cash to be achieved in the digital world. Transferable digital cash systems enable the payee of a transaction to spend the received e-coin in another transaction in which they will play the payer role. In such systems, preventing the double-spending fraud is a hard issue since several people are in possession of the same e-coin throughout its lifetime. The authors of [11] propose a system in which transactions involve the participation of a verifying authority, which checks that e-coins have not been spent before. A recent paper [12] analyzes previous models for transferable digital coins and concludes that they are incomplete. Then, the authors propose a new model and prove its feasibility by giving a concrete construction and rigorous proofs that it satisfies the model.

Modern approaches to digital cash aim to avoid the need for a trusted central entity. This is achieved through the existence of a distributed public ledger storing a record of all the transactions. This is the case for the Bitcoin [13] or the Ethereum [14] cryptocurrencies among many others. Distributed ledgers require the existence of a method for validating transactions without the need to trust a central authority, namely a *consensus mechanism*. The *proof-of-work* approach implemented by Bitcoin has been criticized due to the huge electricity consumption it involves. The *proof-of-stake* alternative, adopted by Ethereum, is much more energy efficient. The design and evaluation of alternative consensus mechanisms is an active area of research.

Attempts to develop a full cash-like digital payment system which is both anonymous, off-line, and secure against double-spending have been forced to include some trusted hardware element. This is the case for the OPERA system [15], which requires a new concept of memory called ORM (one-time-readable memory). The European Commission has reached a similar conclusion to that mentioned in its 'Report on digital euro' [16], which mentions the deployment of two systems in parallel: one based on trusted hardware that can be off-line, anonymous, and without third-party intervention; and an account-based on-line, which is fully software based but excludes the possibility of anonymity.

This paper provides a formal description of a new digital cash paradigm, which enables customers (the payers) to issue no-valued e-coins. These no-valued e-coins are indistinguishable from valued ones and can be used to conduct bogus transactions against the vendor (the payee). The vendor cannot determine whether the received e-coin was valued or not while the customer receives the requested digital item only when a valued e-coin has been spent.

This paradigm has proven its validity in the scope of pay-by-phone parking applications. In [17], a driver, after parking their car in a regulated zone, acquires tickets for some consecutive short time intervals during which their car is expected to be parked. So as to mask the expected parking duration, all the drivers always request the same amount of tickets. Those tickets belonging to intervals after the expected parking duration are paid through no-valued e-coins. An ad-hoc construction belonging to this digital cash paradigm was presented in [17]. Nevertheless, that initial design assumed the use of a fixed suite of cryptosystems due to restrictions in what concerns to the key-size of the used cryptography. More precisely, the cleartexts of the cryptosystem used by the vendor for signing the issued e-coins had to be large enough to accommodate public keys of the cryptosystem used for encrypting the acquired digital item. The role of the mentioned cryptosystems will be explained later in Section 4.

In this paper, we propose how the mentioned key size limitation can be eliminated. The novel construction also avoids the need in [17] for a timestamp authority. Instead, customers timestamp their transactions by themselves.

Section 1 has presented an introduction to digital cash systems. Section 2 briefly reviews the cryptographic tools required in our construction. Next, Section 3 presents a construction based on the use of OAEP (Optimal Asymmetric Encryption Padding [18]), which allows the simulation of digital signatures over messages of arbitrary length. After that, the novel digital cash paradigm is detailed in Section 4. Some cryptosystems providing the required features of the new paradigm are discussed in Section 5. Section 6 is devoted to analyzing the security of the proposal. Next, experimental results are summarized in Section 7, while Section 8 concludes the paper.

2. Preliminaries

This section provides a brief introduction to the cryptographic primitives used by the proposed paradigm.

2.1. Public Key Encryption

In a public key cryptosystem [19], each party is in possession of a private-public key pair. The private key is kept secret while the public one can be made worldwide available. A message encrypted under the public key can only be decrypted by a party in possession of the private one.

2.2. Digital Signatures

Public key cryptography enables the computation of digital signatures [19]. Given M (usually the hash digest of the data to be signed), the signer computes a digital signature over M using their private key. The resulting signature is denoted as $\text{Sign}_{PK}(M)$, and $\{M, \text{Sign}_{PK}(M)\}$ is a digest-signature tuple.

Such a tuple is validated under signer's public key, PK. A positive validation provides integrity, authentication and non-repudiation to the data hashed into M.

2.3. Simulatable Digital Signatures

The term *simulatable* is widely used in the realm of zero-knowledge proofs [20]. A simulator is an entity which, without knowing the secret, can produce a transcript that looks like a proper interaction between a honest prover and the verifier.

In the context of digital signatures, we say a signature scheme is simulatable, when an entity not knowing the secret key is able to produce valid digest-signature tuples. The digest component of a simulated tuple cannot be chosen by the simulator, otherwise such signature scheme would be forgeable. Use of a simulated tuple for authentication purposes further requires finding a piece of data whose digest matches the obtained one. This is unfeasible if an appropriate one-way hash function is being used for digest computation.

2.4. Blind Signatures

Blind signatures [1] are computed through a protocol run between two parties: Alice, who is in possession of a piece of data whose hash digest is M, and Bob who owns a key-pair.

After running the protocol, Alice gets Bob's signature on M, while Bob does not learn any information about M nor about the resulting signature.

3. Message Digests for Simulatable Signatures

As it will be explained next, our digital cash paradigm requires a signature system allowing the computation of simulated digest-signature tuples, which can be linked to a piece of data. So as to make it possible, the one-way hash function employed in traditional digital signature schemes must be replaced with a similar function allowing some degree of pseudo-random reversibility. In this section we propose a construction fulfilling this requirement.

3.1. Optimal Asymmetric Encryption Padding

Optimal Asymmetric Encryption Padding (OAEP) is a procedure initially proposed to pad the plaintext prior to its asymmetric encryption [18]. We next describe OAEP when no plaintext-awareness is required (by setting parameter k_1 of the original proposal to 0). Let m be the bitlength of the input plaintext, and let k_0 be an integer parameter. OAEP is constructed from two of oracles,

$\mathcal{G}$ and $\mathcal{H}$, producing m and k_0 bit outputs, respectively.

The padding procedure takes as input the original plaintext M and a random k_0-bit string r:

1. Compute $X = M \oplus \mathcal{G}(r)$,
2. Compute $Y = r \oplus \mathcal{H}(X)$,
3. Return (X, Y).

We will denote this process as $(X, Y) = \text{OAEP}_{m,k_0}(M, r)$. The resulting X and Y are m and k_0 bits long, respectively. Along the paper, sub-indices m and k_0 will be removed when deemed redundant. The reverse procedure returns the original (M, r) pair from (X, Y):

1. Compute $r = Y \oplus \mathcal{H}(X)$,
2. Compute $M = X \oplus \mathcal{G}(r)$,
3. Return (M, r).

We will denote the reverse process as $(M, r) = \text{OAEP}^{-1}_{m,k_0}(X, Y)$.

The following lemma states a property of OAEP, which is crucial for our construction.

Lemma 1. *Given a message M and a k_0-bit string Y, it is hard to find an $\{r, X\}$ pair so that $(X, Y) = \text{OAEP}_{m,k_0}(M, r)$.*

Proof. Given M and Y, one must find a bitstring r satisfying Equation (1).

$$Y = r \oplus \mathcal{H}(M \oplus \mathcal{G}(r)). \tag{1}$$

Since $\mathcal{G}$ acts as a random oracle, after choosing r, the output of $\mathcal{G}(r)$ is assumed to be random so that $M \oplus \mathcal{G}(r)$ is also random. Function $\mathcal{H}$ is also a random oracle so that $\mathcal{H}(M \oplus \mathcal{G}(r))$ is random and so $r \oplus \mathcal{H}(M \oplus \mathcal{G}(r))$ is. Hence, the probability that the resulting random string matches Y is 2^{-k_0} so that the expected number of trials needed for finding such an r is 2^{k_0} which is unfeasible if k_0 is large enough. Given that r is k_0 bits long, there are exactly 2^{k_0} candidates so that such an r may not exist.

Another possibility is to search for an X satisfying Equation (2).

$$M = X \oplus \mathcal{G}(\mathcal{H}(X) \oplus Y). \tag{2}$$

In this case, an equivalent analysis can be applied leading to a 2^{-m} success probability, which is harder, as typically $m \gg k_0$. □

3.2. Plaintext Awareness

Given a random (X,Y) pair, the result of computing $(M,r) = \text{OAEP}^{-1}_{m,k_0}(X,Y)$ produces a pseudo-random output. A party performing this computation cannot determine whether (X,Y) was generated at random or it was obtained by computing $(X,Y) = \text{OAEP}_{m,k_0}(M,r)$ from an input (M,r) pair. In this latter case, the creator of (X,Y) was aware of plaintext M.

Plaintext awareness is provided by appending k_1 '0' bits to M before running the OAEP process [18]. After running the OAEP reverse procedure, one must check that the obtained message M carries k_1 attached '0' bits which can then be removed. This construction provides plaintext awareness with probability $1 - 2^{-k_1}$ so that a large enough value for k_1 leads to a close to 1 probability.

Throughout this paper, subscript PA will be used to denote that OAEP is being used with the plaintext awareness feature. When plaintext-awareness is provided, string X is k_1 bits longer than M.

3.3. Proposed Construction

We next propose an OAEP-based construction allowing the simulation of digest-signature tuples in such a way that the digest component of simulated tuples can be linked to a piece of data whose length can be chosen by the simulator, but its actual value cannot.

Given an existing simulatable signature scheme, the construction is as follows:

1. Let M be the m-bit message to be signed.
2. Let l be the length of the digests signed by the signature scheme.
3. Generate a random l-bit bitstring r and compute $(X,Y) = \text{OAEP}_{m,l}(M,r)$.
4. Compute a digital signature over Y, namely $\text{Sign}(Y)$.
5. Send the $\{X, Y, \text{Sign}(Y)\}$ tuple to the receiver.

Such a tuple is validated as follows:

1. Validate the $\{Y, \text{Sign}(Y)\}$ digest-signature tuple under signer's public key.
2. Compute $(M, r) = \text{OAEP}^{-1}_{m,l}(X,Y)$ so as to get message M.

Note that a tuple $\{X, Y, \text{Sign}(Y)\}$ can be transformed into $\{X', Y, \text{Sign}(Y)\}$ with $X' \neq X$ while the signature validation still produces a positive result. In such a case, the computation of $(M', r') = \text{OAEP}^{-1}_{m,l}(X', Y)$ leads to a piece of data $M' \neq M$. This is not an issue in our construction as long as M' cannot be chosen by the manipulating party, as it has been stated in Lemma 1. The length of M' corresponds to that of X' (or it is k_1 bits shorter if plaintext-awareness has been set).

Similarly, any party can generate a simulated $\{Y', \text{Sign}(Y')\}$ tuple for the underlying signature scheme and then choose any X' component. The resulting $\{X', Y', \text{Sign}(Y')\}$ tuple will also result in a positive validation. As in the previous remark, this is not an issue since the resulting M' is obtained pseudo-randomly.

The simulatability of the proposal can be disabled by setting the plaintext awareness feature to OAEP. This is because the pseudo-random piece of data M' obtained by simulation does not meet this feature.

4. Novel Digital Cash Paradigm Description

This section provides a detailed description of the proposed digital cash paradigm.

4.1. Overview

Our proposal is a pre-paid digital cash paradigm. Customers acquire valued e-coins in advance and store them in an e-wallet. They will later be spent against the vendor when purchasing items. The paradigm is composed of two actors:

- **Vendor.** A vendor sells digital products online and participates in the issuance of valued e-coins after being paid for them.
- **Customers.** They manage an e-wallet containing valued e-coins. These e-coins are acquired in advance and stored until spent during a purchase procedure. Customers can generate no-valued e-coins on their own.

No-valued e-coins can be spent against the vendor but, in such a case, the customer will not receive any product back. No-valued e-coins enable bogus purchases aiming to mask consumption patterns. The vendor cannot distinguish whether the e-coin involved in a transaction was valued or not.

4.2. e-Coin Composition

Given vendor's public key, PK_V, an e-coin is a tuple of the form represented in Equation (3). Components subindexed with S refer to a cryptographic key-pair used for 'Signing' a transaction. The mentioned key-pair is used by the customer to issue a signature when the e-coin is spent. Those tuple components subindexed with R are related to a key-pair used for 'Receiving' the acquired digital item. Digital items are encrypted by the vendor under public key Q_R before transmitting them to customers.

$$\{v_S, Q_S, (X_S, Y_S), v_R, Q_R, (X_R, Y_R), Y, \text{Sign}_{PK_V}(Y).\} \tag{3}$$

All the e-coin components but the last one ($\text{Sign}_{PK_V}(Y)$) are always generated by the customer. If the e-coin is valued, signature $\text{Sign}_{PK_V}(Y)$ is computed by the vendor; otherwise, it is simulated by the customer. Regarding the components of an e-coin,

- v_S/Q_S is a private/public key-pair of a public key cryptosystem allowing digital signature computation. Hence, data signed with v_s can be validated under Q_S. Q_S has been OAEP_{PA}-encoded (with plaintext-awareness) into $(X_S, Y_S) = \text{OAEP}_{PA}(Q_S, r_S)$ for some random r_S.
- v_R/Q_R is a private/public key-pair of a public key cryptosystem allowing data encryption. Hence, data encrypted under Q_R can only be decrypted by providing v_R. Q_R has been OAEP-encoded into $(X_R, Y_R) = \text{OAEP}(Q_R, r_R)$ for some random r_R.
- Let $Y = Y_S \oplus Y_R$. Then, $\{Y, \text{Sign}_{PK_V}(Y)\}$ is a digest-signature tuple which can be validated under PK_V.

4.3. Valued e-Coin Generation

A valued e-coin is generated through a procedure in which both the customer and the vendor do participate.

1. The customer pays the vendor the price of an e-coin.
2. The customer generates a random private key v_S and the corresponding public one Q_S. The customer also generates a random r_S and computes $(X_S, Y_S) = \text{OAEP}_{PA}(Q_S, r_S)$.
3. The customer generates a random private key v_R and the corresponding public one Q_R, and computes $(X_R, Y_R) = \text{OAEP}(Q_R, r_R)$ for some random r_R chosen by the customer.
4. The customer computes $Y = Y_S \oplus Y_R$.
5. The customer requests the vendor to compute a blind signature on Y. Let $\text{Sign}_{PK_V}(Y)$ be the resulting signature. Hence, $\{Y, \text{Sign}_{PK_V}(Y)\}$ is a digest-signature tuple.

At the end of this process, the customer is in possession of an e-coin tuple as that shown in Equation (3).

4.4. No-Valued e-Coin Generation

A no-valued e-coin is generated by the customer on their own.

1. The customer generates a simulated message-signature tuple under vendor's public key. Let $\{Y, \text{Sign}_{PK_V}(Y)\}$ be the simulated tuple.
2. The customer generates a random private key v_S and the corresponding public one Q_S. The customer also generates a random r_S and computes $(X_S, Y_S) = \text{OAEP}_{PA}(Q_S, r_S)$.
3. The customer calculates $Y_R = Y \oplus Y_S$, generates a random X_R, and computes $(Q_R, r_R) = \text{OAEP}^{-1}(X_R, Y_R)$. If Q_R is not a valid public key, this step is run again taking a different X_R.

At the end of this process, the customer is in possession of a partial e-coin tuple as that shown in Equation (4).

$$\{v_S, Q_S, (X_S, Y_S), \varnothing, Q_R, (X_R, Y_R), Y, \text{Sign}_{PK_V}(Y)\}. \tag{4}$$

The private key v_R corresponding to Q_R is not known, and the corresponding part of the tuple is empty ($\varnothing$).

Note that we need a cryptosystem in which the probability of obtaining a valid public key in a pseudo-random manner is relatively high (step 3). More details are given in Section 5.3.

4.5. Spending an e-Coin

A customer wishing to acquire some product P whose price is worth the value of an e-coin asks the vendor to engage in the following procedure:

1. The customer sends $\{(X_S, Y_S), (X_R, Y_R), \text{Sign}_{PK_V}(Y)\}$ to the vendor together with a digital signature $\text{Sign}_{Q_S}(\mathcal{H}(CurrentTime||Y_S||Y_R))$ computed with private key v_S ($\mathcal{H}$ is a hash function).
2. The vendor runs $(Q_S, r_S) = \text{OAEP}_{PA}^{-1}(X_S, Y_S)$. If the plaintext-awareness checking is met, they check the digital signature received at the previous step under Q_S. In case of failure, the e-coin is rejected.
3. The vendor computes $Y = Y_S \oplus Y_R$ and checks that $\{Y, \text{Sign}_{PK_V}(Y)\}$ is a valid digest-signature tuple under vendor's public key PK_V.
4. The vendor checks that no e-coin with the same Y_S component has been spent before. In such a case, the previously stored digital signature, which includes the time it was spent for the first time, is returned as a proof of double spending and the transaction is rejected. Otherwise, all the data received at step 1 is stored by the vendor.
5. The vendor computes $(Q_R, r_R) = \text{OAEP}^{-1}(X_R, Y_R)$.
6. The vendor encrypts the product P under public key Q_R (creating a digital envelope if P is large) and sends the resulting ciphertext to the customer.
7. If the spent e-coin was valued, the customer decrypts the received ciphertext using private key v_R, getting P as a result. Otherwise, this step is skipped and the customer does not get any product.

5. Cryptosystems Choice

This section provides an assessment on the features to be provided by the cryptosystems chosen to implement the paradigm.

5.1. Cryptosystem for Vendor'S Key-Pair

In the described digital cash paradigm, the vendor is in possession of a key-pair whose public key has been denoted as PK_V. This key-pair is used for the generation of the $\{Y, \text{Sign}_{PK_V}(Y)\}$ digest-signature tuple. This tuple is generated differently depending on whether the generated e-coin is valued or not. More precisely,

- If the e-coin is valued, the customer computes Y and requires the vendor to compute a blind signature on it (Section 4.3, step 5).

- If the e-coin is no-valued, the tuple is simulated by the customer. The vendor does not take part in this process (Section 4.4, step 1).

Hence, the signature scheme chosen for such $\{Y, \text{Sign}_{PK_V}(Y)\}$ tuples has to enable both:

- The computation of *blind* signatures.
- The generation of *simulated* digest-signature tuples.

Next, we discuss two feasible options.

5.1.1. RSA Signatures

Given an RSA [21] private key, d, and the corresponding public one (N, e), an RSA digital signature over M is computed from secret key d as $S = M^d \pmod{N}$. The resulting $\{M, S\}$ digest-signature tuple is verified by checking whether M equals $S^e \pmod{N}$.

RSA blind signatures [1] can be issued through the following protocol:

1. Alice chooses a random $R \in \mathbb{Z}_N$ and computes $\overline{M} = M \cdot R^e \pmod{N}$ and sends $\overline{M}$ to Bob (operator · denotes the integer modular multiplication).
2. Bob computes $\overline{S} = \overline{M}^d \pmod{N}$ and sends $\overline{S}$ to Alice.
3. Alice computes $S = \overline{S} \cdot R^{-1} \pmod{N}$ obtaining signature S on M.

RSA digest-signature tuples can be simulated by taking a random $S \in \mathbb{Z}_N$ and then computing $M = S^e \pmod{N}$. In a typical signature, M is the hash digest of the piece of data to be signed. Hence, obtaining a message signed by simulation further requires inversion of such hash function which is unfeasible.

We can enable RSA signatures with the construction presented in Section 3.3. By signing the Y component of an OAEP-encoded message, we allow simulated signatures over pseudo-random pieces of data.

5.1.2. Boldyreva Signatures

Boldyreva digital signatures [22] are discrete-logarithm based and implemented over a so-called *Gap Diffie–Hellman* (GDH) group. In a GDH group, the *Diffie–Hellman problem* is difficult, namely, given g^a and g^b, computing g^{ab} is assumed to be hard. However, the *Decisional Diffie–Hellman problem* is easy to solve, namely, given g^a, g^b, and g^c, it is easy to decide whether $c = ab$.

A GDH group of large prime order q has to be chosen. Let g be a generator of such group. A private key is generated by choosing a random $x \in \{0, \ldots, q-1\}$. The corresponding public key is computed as $y = g^x$.

A digital signature over a digest M is computed as $S = M^x$.

A digest-signature tuple $\{M, S\}$ is validated under public key y by checking whether $\{M, y, S\}$ is a GDH-tuple, that is, $\log_g M \cdot \log_g y = \log_g S$.

This signature scheme allows the computation of blind signatures throughout the following procedure:

1. Alice chooses $r \in \{0, \ldots, q-1\}$ and computes $\overline{M} = M \cdot g^r$. Then she sends $\overline{M}$ to Bob.
2. Bob computes $\overline{S} = \overline{M}^x$ and sends $\overline{\sigma}$ back to Alice.
3. Finally, Alice computes $S = \overline{S} \cdot y^{-r}$ which is a digital signature over M.

Given the group generator g and a public key y ($y = g^x$), a Boldyreva digest-signature tuple is simulated by taking a random integer t and computing $M = g^t$ and $S = y^t$. Tuple $\{M, S\}$ is a simulated digest-signature tuple. The construction in Section 3.3 allows us to link simulated signatures to pseudo-random pieces of data.

5.2. Cryptosystem for e-Coin Transaction Signature

When a customer spends an e-coin (Section 4.5, step 1) they make use of the v_S private key to issue a digital signature that will be validated under Q_S. This $\{v_S, Q_S\}$ key-pair is always generated through the traditional (private key first, public key next) procedure. Hence, any cryptosystem allowing digital signature computation can be chosen.

5.3. Cryptosystem for Product Encryption

When the customer spends an e-coin to acquire some product, the vendor encrypts it under the Q_R public key included in the spent e-coin (Section 4.5, step 6). The customer is only able to decrypt such ciphertext if they know the corresponding private key. This public key is generated differently depending on whether the e-coin it is contained in is valued or not.

- If the e-coin is valued, the customer generates private key v_R and then the corresponding public one Q_R (Section 4.3, step 3).
- If the e-coin is no-valued, public key Q_R is obtained pseudo-randomly (Section 4.4, step 3).

Therefore, the cryptosystem for such $\{v_R, Q_R\}$ key-pairs must satisfy the following requirements:

- It allows public key data encryption;
- It provides a relatively high probability of obtaining a valid public key by means of a pseudo-random process;
- It cannot be determined whether a given public key has been generated together with its private counterpart (Section 4.3 step 3) or through a pseudo-random process (Section 4.4, step 3).

The RSA [21] cryptosystem would not be a suitable option since the probability that a pseudo-random integer N is composed of two large prime factors is rather low.

We next detail two suitable public key encryption schemes.

5.3.1. ECIES

The Elliptic Curve Integrated Encryption Scheme (ECIES) [23] is an elliptic curve-based public key encryption scheme whose security holds on the assumed intractability of the *Elliptic Curve Discrete Logarithm Problem* (ECDLP).

Such a cryptosystem is set by choosing an elliptic curve E represented as an expression of the form shown in Equation (5):

$$Y^2 = X^3 + AX + B, \qquad (5)$$

with A, B being elements of a finite field $\mathbb{F}$ such that its set of points $E(\mathbb{F})$ has a cardinality divisible by a large prime q. An order-q point P of $E(\mathbb{F})$ is also chosen. Throughout this section, we assume q is prime.

An ECIES private key is generated by choosing a random $v \in \{0, \ldots, q-1\}$. The corresponding public key is the point of $E(\mathbb{F})$ computed as $Q = vP$.

The probability that a random point $(x, y) \in \mathbb{F} \times \mathbb{F}$ belongs to $E(\mathbb{F})$ is negligible since its components should satisfy Equation (5). Nevertheless, this drawback can be addressed by representing elliptic curve points in compressed form. A point $(x, y) \in E(\mathbb{F})$ can be represented as (x, b) with b being a Boolean indicating whether $y > -y$. In this way, a randomly generated compressed point (x, b) belongs to $E(\mathbb{F})$ (and hence it is a public key), if its x-component satisfies that $x^3 + Ax + B$ is a quadratic residue in $\mathbb{F}$. This happens with a close to $1/2$ probability [17].

5.3.2. ElGamal

ElGamal [24] is a public key cryptosystem whose security holds on the assumed intractability of the *Discrete Logarithm Problem* (DLP).

This cryptosystem is set by choosing a large prime q satisfying that $p = 2q + 1$ is also prime. The cryptosystem is built on the order-q multiplicative subgroup of $\mathbb{Z}_p^*$. An order-q element g is chosen during the setup.

A private key is generated by choosing a random $x \in \{0, \ldots, q-1\}$ and the corresponding public key is computed as $y = g^x \pmod{p}$.

A randomly selected element from $\mathbb{Z}_p^*$ turns out to be a public key if its order is q. This happens exactly with a 1/2 probability since $\mathbb{Z}_p^*$ contains exactly $p-1 = 2q$ elements with q of them having the desired order.

6. Security Analysis

A digital cash system like the one presented in this paper should satisfy the following security requirements:

1. Valued e-coins cannot be forged by malicious customers;
2. E-coins cannot be double-spent;
3. Customers cannot be falsely accused of double-spending an e-coin.

Although the vendor can be assumed to be a somehow trusted party, Req. 3 is still needed to prevent malicious double-spenders from claiming they are being accused falsely. The following lemmas address the fulfillment of the enumerated requirements.

Lemma 2. *Valued e-coins of the proposed digital cash paradigm cannot be forged.*

Proof. Let us recall that an e-coin is a tuple of the form shown in Equation (3). An e-coin can only be spent if private key v_S is known. Otherwise, the digital signature required at step 1 of the "Spending" protocol cannot be computed. Hence, Q_S must be generated together with v_S, and the Y_S component of its OAEP encoding (with plaintext awareness) is obtained pseudo-randomly by calling $\text{OAEP}_{PA}(Q_S, r_S)$ for some random r_S. Hence, the Y_S component cannot be chosen by a dishonest party aiming to forge an e-coin. Note also that the Y_S component of spent e-coins is checked not to be part of an already spent e-coin. In this way, there is no point in taking the Y_S component of a new e-coin from an existing one.

If the forger then simulates the $\{Y, \text{Sign}_{PK_V}(Y)\}$ digest-signature pair, the resulting Y cannot be chosen (otherwise the underlying signature scheme would be forgeable), so that $Y_R = Y \oplus Y_S$ can neither be chosen and, after taking any X_R, the public key Q_R obtained from $(Q_R, r_R) = \text{OAEP}^{-1}(X_R, Y_R)$ is pseudo-random and its private key remains unknown. In this way, the resulting e-coin is no-valued. Lemma 1 guarantees that given Y_R and some chosen Q_R, finding a $\{r_R, X_R\}$ pair satisfying the previous expression is unfeasible.

Alternatively, the forger could generate a v_R/Q_R key-pair and OAEP-encode it into (X_R, Y_R). In this case, the obtained Y_R component is pseudo-random and so the resulting $Y = Y_S \oplus Y_R$ is. Hence, the signature $\text{Sign}_{PK_V}(Y)$ over Y cannot be obtained by the forger without the participation of the vendor. □

Lemma 3. *E-coins of the proposed digital cash system cannot be double-spent.*

Proof. When an e-coin is spent, the vendor stores a record which includes its Y_S component. Hence, any attempt to spend the same e-coin in the future will be detected. □

Lemma 4. *An honest customer cannot be falsely accused of being a double-spender by a dishonest vendor.*

Proof. Customers spending an e-coin are required to digitally sign a timestamped sequence using the v_S private key. This digital signature can be validated under public key Q_S. Only the customer who generated an e-coin knows its v_S secret key.

A vendor claiming that an e-coin is being double-spent is required to provide the signed timestamped sequence of the first time the e-coin was spent (Section 4.5, step 4). If the claim is false, they will be unable to provide it. □

7. Experimental Results

The proposed paradigm has been validated through a prototype implemented in Java. Cryptographic operations involving large integers use the `java.math.BigInteger` library. Hash digests have been computed using the SHA-224 [25] function. Regarding the employed cryptosystems, we have chosen the following:

- *Vendor's key-pair (Section 5.1)*: RSA with 2048 bit keys.
- *Cryptosystem for e-coin transaction signature (Section 5.2)*: ECDSA [26] with 224 bit keys.
- *Cryptosystem for product encryption (Section 5.3)*: ECIES with 224 bit keys.

Our experiments have measured the running time of the "Valued e-coin generation" (Section 4.3), "No-valued e-coin generation" (Section 4.4), and "Spending an e-coin" (Section 4.5) procedures. The prototype has been run on several personal computers. Average running times from 500 executions have been measured. As expected, computers with a faster processor lead to better running times. We have also observed that the running time benefits from parallel execution mode.

Table 1 shows the average running time of the "Valued e-coin generation" and "No-valued e-coin generation" procedures. Let us recall that the generation of valued e-coins involves both the customer and the server (which is required to compute a blind signature) while the procedure for generating no-valued ones is run entirely by the customer. The table shows that the generation of a no-valued e-coin takes some more time than a valued one. This is due to the fact that step 3 of the procedure for generating no-valued e-coins sometimes has to be run more than one time. In our experiments, in which we have implemented the ECIES cryptosystem, there is a 50% chance of having to run it again. In the fastest tested processor, in parallel mode, generation of a valued e-coin and a no-valued one takes around 3 and 4 ms, respectively, leading to generation rates of 333 and 250 e-coins per second, respectively.

Table 1. "E-coin generation" running times (in milliseconds).

Processor	System Server & Client			Valued e-Coin		No-Valued e-Coin	
	Cores	Threads	GHz	Serial	Parallel	Serial	Parallel
AMD Athlon	4	4	2.80	49.28	13.62	68.25	20.39
Intel i5-8350U	4	8	1.70–3.60	21.40	5.23	32.69	7.69
Intel i7-6700	4	8	3.40–4.00	20.50	4.71	28.95	7.27
Intel i7-8700	6	12	3.20–4.60	18.75	4.66	28.41	6.31
AMD Ryzen 7	8	16	3.70–4.30	23.18	3.05	32.86	3.97

Table 2 shows the running time of the "Spending an e-coin" procedure at the vendor part. We focus on this part of the process due to the fact that a vendor may receive a lot of concurrent payments. We do not distinguish between spending a valued or a no-valued e-coin since the procedure is exactly the same in both cases. The fastest running time, obtained on an AMD Ryzen 7 processor in parallel mode, indicates that receiving an e-coin payment takes 2.69 ms, so that around 371 payments can be processed in a single second.

Table 2. "Spending an e-coin" running times (in milliseconds).

Processor	System Server				
	Cores	Threads	GHz	Serial	Parallel
AMD Athlon	4	4	2.80	51.24	13.99
Intel i5-8350U	4	8	1.70-3.60	26.67	7.37
Intel i7-6700	4	8	3.40-4.00	22.68	5.12
Intel i7-8700	6	12	3.20-4.60	20.69	3.59
AMD Ryzen 7	8	16	3.70-4.30	25.41	2.69

8. Conclusions

This paper has presented a novel digital cash paradigm in which customers are able to generate no-valued e-coins by themselves. Such no-valued e-coins can be spent like regular valued ones in such a way that the vendor receiving a payment is unable to distinguish between both situations. The customer only receives the requested digital product when the spent e-coin is a valued one. This new paradigm fits in scenarios in which customers may wish to mask their consumption patterns through bogus transactions like pay-per-view TV or music platforms. The paradigm has already proven its validity in privacy-preserving pay-by-phone parking systems enabling drivers with the possibility of keeping their expected parking time secret.

In our future research, we plan to investigate the design of privacy-preserving protocols, which include the presented digital cash paradigm as a building block.

Author Contributions: Conceptualization, R.B. and F.S.; methodology, R.B. and F.S.; validation, R.B. and F.S.; formal analysis, F.S.; writing—original draft preparation, F.S.; writing—review and editing, R.B.; funding acquisition, F.S. All authors have read and agreed to the published version of the manuscript.

Funding: This research was funded by the Spanish Ministry of Science, Innovation and Universities grant number MTM2017-83271-R.

Institutional Review Board Statement: Not applicable.

Informed Consent Statement: Not applicable.

Data Availability Statement: Data is contained within the article.

Conflicts of Interest: The authors declare no conflict of interest. The funders had no role in the design of the study; in the collection, analyses, or interpretation of data; in the writing of the manuscript, or in the decision to publish the results.

References

1. Chaum, D. Blind Signatures for Untraceable Payments. In *Advances in Cryptology*; Chaum, D., Rivest, R.L., Sherman, A.T., Eds.; Springer: Boston, MA, USA, 1983; pp. 199–203.
2. Brands, S. Untraceable Off-line Cash in Wallet with Observers. In *Advances in Cryptology—CRYPTO'93*; Stinson, D.R., Ed.; Springer: Berlin/Heidelberg, Germany, 1994; pp. 302–318.
3. Eng, T.; Okamoto, T. Single-term divisible electronic coins. In *Advances in Cryptology—EUROCRYPT'94*; De Santis, A., Ed.; Springer: Berlin/Heidelberg, Germany, 1995; pp. 306–319.
4. Nakanishi, T.; Sugiyama, Y. Unlinkable Divisible Electronic Cash. In *Information Security*; Goos, G., Hartmanis, J., van Leeuwen, J., Pieprzyk, J., Seberry, J., Okamoto, E., Eds.; Springer: Berlin/Heidelberg, Germany, 2000; pp. 121–134.
5. Canard, S.; Gouget, A. Divisible E-Cash Systems Can Be Truly Anonymous. In *Advances in Cryptology—EUROCRYPT 2007*; Naor, M., Ed.; Springer: Berlin/Heidelberg, Germany, 2007; pp. 482–497.
6. Au, M.H.; Susilo, W.; Mu, Y. Practical Anonymous Divisible E-Cash from Bounded Accumulators. In *Financial Cryptography and Data Security*; Tsudik, G., Ed.; Springer: Berlin/Heidelberg, Germany, 2008; pp. 287–301.
7. Liu, J. Efficient Arbitrarily Divisible E-Cash Applicable to Secure Massive Transactions. *IEEE Access* **2019**, *7*, 59299–59310. [CrossRef]
8. Bourse, F.; Pointcheval, D.; Sanders, O. Divisible E-Cash from Constrained Pseudo-Random Functions. In *Advances in Cryptology—ASIACRYPT 2019*; Galbraith, S.D., Moriai, S., Eds.; Springer International Publishing: Cham, Switzerland, 2019; pp. 679–708.
9. Rivest, R.L.; Shamir, A. PayWord and MicroMint: Two simple micropayment schemes. In *International Workshop on Security Protocols*; Springer: Berlin/Heidelberg, Germany, 1996; pp. 69–87.
10. Oros, H.; Popescu, C. A Secure and Efficient Off-Line Electronic Payment System for Wireless Networks. *Int. J. Comput. Commun. Control.* **2010**, *V*, 551–557. [CrossRef]
11. Sai Anand, R.; Madhavan, C. An Online, Transferable E-Cash Payment System. In *Progress in Cryptology —INDOCRYPT 2000*; Roy, B., Okamoto, E., Eds.; Springer: Berlin/Heidelberg, Germany, 2000; pp. 93–103.
12. Bauer, B.; Fuchsbauer, G.; Qian, C. Transferable E-Cash: A Cleaner Model and the First Practical Instantiation. In *Public-Key Cryptography—PKC 2021*; Garay, J.A., Ed.; Springer International Publishing: Cham, Switzerland, 2021; pp. 559–590.
13. Nakamoto, S. Bitcoin: A Peer-to-Peer Electronic Cash System. 2009. pp. 1–9. Available online: https://bitcoin.org/bitcoin.pdf (accessed on 22 September 2021).
14. Wood, G. Ethereum: A secure decentralised generalised transaction ledger. *Ethereum Proj. Yellow Pap.* **2021**, *151*, 1–32.
15. Park, K.W.; Baek, S.H. OPERA: A Complete Offline and Anonymous Digital Cash Transaction System with a One-Time Readable Memory. *IEICE Trans. Inf. Syst.* **2017**, *100*, 2348–2356. [CrossRef]
16. European Central Bank. *Report on Digital Euro*; Tech. Report; Frankfurt am Main, Germany, 2020. Available online: https://www.ecb.europa.eu/pub/pdf/other/Report_on_a_digital_euro~4d7268b458.en.pdf (accessed on 22 September 2021).
17. Borges, R.; Sebé, F. An efficient privacy-preserving pay-by-phone system for regulated parking areas. *Int. J. Inf. Secur.* **2021**, *20*, 715–727. [CrossRef]

18. Bellare, M.; Rogaway, P. *Optimal Asymmetric Encryption—How to Encrypt with RSA*; Springer: Berlin/Heidelberg, Germany, 1995; pp. 92–111.
19. Schneier, B. *Applied Cryptography: Protocols, Algorithms, and Source Code in C*, 2nd ed.; John Wiley & Sons, Inc.: Hoboken, NJ, USA, 1995.
20. Goldwasser, S.; Micali, S.; Rackoff, C. The knowledge complexity of interactive proof systems. *SIAM J. Comput.* **1989**, *18*, 186–208. [CrossRef]
21. Rivest, R.L.; Shamir, A.; Adleman, L. A Method for Obtaining Digital Signatures and Public-Key Cryptosystems. *Commun. ACM* **1978**, *21*, 120–126. [CrossRef]
22. Boldyreva, A. Threshold Signatures, Multisignatures and Blind Signatures Based on the Gap-Diffie-Hellman-Group Signature Scheme. In *Public Key Cryptography—PKC 2003*; Desmedt, Y.G., Ed.; Springer: Berlin/Heidelberg, Germany, 2002; pp. 31–46.
23. Gayoso, V.; Hernandez, L.; Sánchez, C. A Survey of the Elliptic Curve Integrated Encryption Scheme. *J. Comput. Sci. Eng.* **2010**, *2*, 7–13.
24. ElGamal, T. A public key cryptosystem and a signature scheme based on discrete logarithms. *IEEE Trans. Inf. Theory* **1985**, *31*, 469–472. [CrossRef]
25. Handschuh, H. SHA Family (Secure Hash Algorithm). In *Encyclopedia of Cryptography and Security*; van Tilborg, H.C.A., Ed.; Springer: Boston, MA, USA, 2005; pp. 565–567. [CrossRef]
26. Johnson, D.; Menezes, A.; Vanstone, S.A. The Elliptic Curve Digital Signature Algorithm (ECDSA). *Int. J. Inf. Secur.* **2001**, *1*, 36–63. [CrossRef]

Article

Authorization Mechanism Based on Blockchain Technology for Protecting Museum-Digital Property Rights

Yun-Ciao Wang [1], Chin-Ling Chen [2,3,4,*] and Yong-Yuan Deng [2,*]

1. National Museum of Marine Biology and Aquarium, Pingtung 94450, Taiwan; yunciao@nmmba.gov.tw
2. Department of Computer Science and Information Engineering, Chaoyang University of Technology, Taichung 41349, Taiwan
3. School of Information Engineering, Changchun Sci-Tech University, Changchun 130600, China
4. School of Computer and Information Engineering, Xiamen University of Technology, Xiamen 361005, China
* Correspondence: clc@mail.cyut.edu.tw (C.-L.C.); allen.nubi@gmail.com (Y.-Y.D.)

Featured Application: Museums not only achieve the goal of promoting social education, but also solve their financial problems.

Abstract: In addition to the exhibition, collection, research, and educational functions of the museum, the development of a future museum includes the trend of leisure and sightseeing. Although the museum is a non-profit organization, if it can provide digital exhibits and collections under the premises of "intellectual property rights" and "cultural assets protection", and licensing and adding value in various fields, it can generate revenue from digital licensing and handle the expenses of museum operations. This will be a new trend in the sustainable development of museum operations. Especially since the outbreak of COVID-19 at the beginning of this year (2020), the American Alliance of Museums (AAM) recently stated that nearly a third of the museums in the United States may be permanently closed since museum operations are facing "extreme financial difficulties." This research is aimed at museums using the business model of "digital authorization". It proposes an authorization mechanism based on blockchain technology protecting the museums' digital rights in the business model and the application of cryptography. The signature and time stamp mechanism achieve non-repudiation and timeless mechanism, which combines blockchain and smart contracts to achieve verifiability, un-forgery, decentralization, and traceability, as well as the non-repudiation of the issue of cash flow with signatures and digital certificates, for the digital rights of museums in business. The business model proposes achievable sustainable development. Museums not only achieve the goal of promoting social education, but also solve their financial problems.

Keywords: museum; digital copyright management; blockchain; smart contract; authorization model

Citation: Wang, Y.-C.; Chen, C.-L.; Deng, Y.-Y. Authorization Mechanism Based on Blockchain Technology for Protecting Museum-Digital Property Rights. *Appl. Sci.* **2021**, *11*, 1085. https://doi.org/10.3390/app11031085

Received: 23 December 2020
Accepted: 21 January 2021
Published: 25 January 2021

Publisher's Note: MDPI stays neutral with regard to jurisdictional claims in published maps and institutional affiliations.

Copyright: © 2021 by the authors. Licensee MDPI, Basel, Switzerland. This article is an open access article distributed under the terms and conditions of the Creative Commons Attribution (CC BY) license (https://creativecommons.org/licenses/by/4.0/).

1. Introduction

In addition to their exhibition, collection, research, and education functions, museums' main purpose is to display and protect cultural resources. Continuous attention has been paid to them. However, there is a difficulty: If these collections are displayed in public places for a long time, they may deteriorate. On the other hand, if they are kept in a warehouse, visitors cannot share this valuable information. In 2007, Ross Parry suggested that the concept of digital collections should be added to the main concepts of museums [1]. The main purpose is to digitize these collections. Moreover, transforming collections into digital content in a unified format and developing them into good digital rights management will not only help promote social education, but also facilitate the operation of museums.

Museum digitalization means that the museum converts the texts, images, and videotapes through digital scanners and digital cameras based on the collections of the museum to produce digital data that can be processed by a computer; "Digital Collection" refers to the data and files on various utensils, paintings, and calligraphy, specimens, and documents that have been processed through digital processes. The "digital collections" authorization originated at the beginning of photography in the 19th century. The British Museum accepted donations of photographic images, as well as professional photographers' photo collections, and sold the collections taken in the museum and the records of museum activity photos; this was the beginning of the museum's image recording and image authorization [2].

In the 20th century, international museums and governments implemented the digitization plan of various museum collections based on the mission of collection preservation and promotion of cultural policies. Currently, in the 21st century, digital technology is booming, and museums have entered the era of digitization. There are a huge number of digital images. Production allows museums to hold numerous copyrights, signaling an important turning point for image authorization. The international museum community has invested a lot of money and human resources starting more than ten years ago to digitize its collections on a large scale. For example, J. Paul Getty Trust, associated with the Getty Museum, paid US$4.2 million from 1997 to 2002 in funding to establish the "Electronic Cataloguing Initiative", which sponsored 21 Los Angeles area museums whose main collections are visual arts.

In 2009, the foundation launched OSCI in cooperation with the J. Paul Getty Museum and eight other institutions. Arthur M. Sackler and Freer Art Gallery; Los Angeles County Museum of Art; National Gallery of Art in Washington, DC; San Francisco Museum of Modern Art; Seattle Art Museum; and the Tate and Walker Art Center. The goal of the alliance is to create models for online catalogs, which will greatly increase access to museum collections to provide interdisciplinarity and the latest research, and innovate how to conduct, introduce, and use this research [3]. In 2002, the Culture Online Project of the British Department of Culture, Media and Sports was founded [4]; the British Museum established the "Merlin Project" in 2006 along with other projects, which are all efforts related to the museum's digital collection.

The core of the museum is its collection and heritage, the physical evidence of human survival and its environment. This includes two levels of connotation: One is the cultural relic entity in the museum's collection; the other is the information resources that recreate the cultural relic entity, reveal its original information and cultural connotation, including text introductions, images, video three-dimensional models, etc. Museum experts and scholars research and publish works on a certain collection or collection preservation technology, as well as works of collection pictures taken by museum photographers, etc., all belonging to the collection resources.

The "Creative Economy Report 2010" of UNESCO [4] points out that "cultural heritage" is the source of all art forms, and the soul of culture and creative industries, which brings together history, anthropology, ethnology, aesthetics, and social perspectives, while influences people's creativity. The intellectual property authorization of the museum means that the museum authorizes the copyright of its collection resources. It includes cultural relics, specimens, and artworks to other institutions for the development of cultural derivatives, transforms cultural resources into cultural goods, and establishes effective communication with consumers. It forms a unique brand of museums and reflects the intention of museums to develop products [5]. The authorized person pays the corresponding fee to the authorizer, and the authorizer gives the authorized person corresponding guidance and assistance. In particular, museums in various countries with rich collections can serve as models for brand authorization.

Brand authorization began in the United States in the early 20th century. When Disney's classic cartoon image of Mickey Mouse became famous, a furniture merchant paid Walt Disney US$300 in exchange for the right to print the image of Mickey Mouse on its products. Disney is recognized as the originator of international brand authorization. Currently, brand authorization has become a global industry with a relatively mature operation model and a complete industrial chain. According to the "2019 Global Licensing Industry

Market Survey Report" released by the International Licensing Industry Merchandiser's Association (LIMA) [6], the global retail sales of licensed goods reached 280.3 billion U.S. dollars in 2018, a year-on-year increase of 3.2%. Among the competitors, China's authorized industry market sales reached 9.5 billion U.S. dollars, maintaining a rapid growth trend with an increase of 67%.

As an image producer, the core mission of museums is to produce images in the spirit of equality, sharing, and reciprocity. This view also echoes the concept of equality of museums. In the comprehensive digital collection, most of the collections that cannot be displayed or watched in permanent exhibitions or special exhibitions can have the opportunity to be presented to the world. For example, the sea area around the National Museum of Marine Biology and Aquarium, located in the Kenting National Park in southern Taiwan, is a typical marine environment intersection, covering the estuary area, sandy mud bottom, reef shores, and other habitats. Chang et al. [7] studied and integrated the fish species in the sub-tidal zone around the National Museum of Marine Biology and Aquarium, which provides a constant monitoring and conservation research platform for the aquatic environment and biodiversity. The museum has also carried out the image management collection of collection resources [8], but how to use these valuable research resources of the museum through the appropriate preservation, management, authorization, and promote social education is an extremely important challenge.

In recent years, under the concept of "activating and reproducing collections", museums spread a huge amount of knowledge and culture to visitors with their rich collections, such as artworks, crafts, biological specimens, texts, drawings, paintings, photos, maps, movies, and sound recordings. Museums all over the world take marketization, digitization, diversification, and popularization as their development direction. Their development and utilization of digital image resources in the collections, via different authorization models, are widely praised by society.

The cultural industry chain is divided into four links: Research and development, production, circulation, and consumption. With the development over time, the term "authorization" has been widely used in the cultural industry, and its connotation and extension have also been continuously expanded, and gradually valued by museums. At present, there are two views on the definition of digital image authorization of museum collections: One view is that authorization refers to the process by which the museum grants the digital image of cultural relics owned or managed by the museum as the subject matter to the authorized person in the form of a contract; another view is that authorization is mainly the process of transaction and management of related intellectual property rights. The ultimate goal of museums' digital authorization of collections is to increase economic benefits based on spreading culture and exerting its educational function.

However, due to various reasons, most people may not be able to visit their favorite museums one by one due to time and space constraints. For example, since the outbreak of the COVID-19 at the beginning of this year (2020), the American Alliance of Museums (AAM) recently stated that nearly one-third of museums in the United States may be permanently closed, and pointed out that museum operations are facing "extreme financial difficulties" [9]. Therefore, determining how to protect museum collections and effectively use these collection resources to maintain the operation of the museum is a critical topic for consideration by museum operators.

Due to the fading of museum collections, while promoting social education, we must strive to preserve them. Digitizing collections is a feasible way. On the other hand, in order to maintain the sustainable operation of museums, it is important to manage the property rights of museum collections after digitization. Copyright provides a bridge between art and commerce because we need to protect the collections. In the past, using watermarking technology to achieve digital property management has been a mature technology [10–12]. Digital rights management is always inseparable from cryptographic technology [13–16]. Up to now, watermarks are combined with smart contract technology to realize digital property management [17]. In recent years, more scholars have used the characteristics of decentralization, non-tampering, traceability, and blockchain openness to solve the application problems of digital rights management, a process that has expanded rapidly [17–22].

However, none of the above-mentioned digital property rights management mechanisms integrate the operation of the cash flow system, and naturally cannot reflect its feasibility. Therefore, this article integrates cash flow management into our digital rights management regarding comprehensive digital collections and promotes transparency of collections, the heart of museums. Apart from the practice of equality, the production of images provides an extension of museum collections and serves as a carrier of culture. The circulation of copied images creates richer and more diverse ways of use [23].

In 2017, Ma proposed a common, flexible, and extendable solution for variant DRM scenes, and can support rapid and customized development [24]. Du Toit proposed a decentralized architectural model, which makes use of digital rights management to enforce access control over personal information [25]. Mrabet et al. [26] concluded the open research issues and future directions towards securing IoT. Including the use of blockchain to address security challenges in IoT, and the implications of IoT deployment in 5G and beyond. Therefore, the first focus of digital rights management is how to achieve proper authorization. Generally, the authorization mode of digital collections in museums is divided into the following three methods:

1. Direct authorization model of museum digitized collections

The direct authorization model is a model in which the museum, as the authorized party, signs a contract with the authorized party to authorize it to use the digital resources of the collections. The museum collects cultural relics, produces digital content, encrypts and encapsulates, authorizes the identity verification and makes remittance notices, authorizes remittances royalties' feedback, and finally operates the key authorization process. The authorization model process is shown in Figure 1. The National Museum of the Netherlands and the British Museum, as well as the National Palace Museum in Taipei in Taiwan, are typical examples of the direct authorization model.

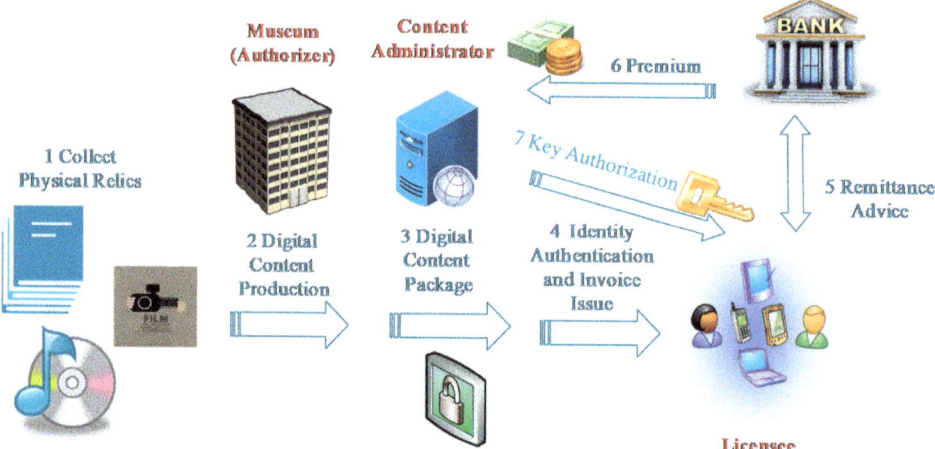

Figure 1. The digital direct authorization model of museum collections.

Under this authorization model, the authorized party often directly participates in the use of the digital image resources of cultural relics by third-party manufacturers. The advantage is that it is not only conducive to the museum as the authorized party to promptly understand the development of digital image resources, but is also given an in-depth understanding of the connotations of the collection by the relevant departments of the museum, which is often helpful to the successful development of digital resources. However, the shortcomings of this authorization model are also obvious. Because the authorized party is a state-owned museum, the nature of its public welfare institutions often makes it limited in authorization methods, scope, personnel incentives, and so on, so it can easily lead to insufficient responses to market demand and changes.

2 Proxy authorization model of museum digitized collections

The proxy authorization model refers to the model in which the museum does not directly act as the authorized subject, but entrusts an agent or an authorization platform as an intermediary, authorizes through a contract with the authorized party, and finally uses the digital resources of the collection in the manner agreed to in the contract. In this model, there will be two authorization behaviors: The first time is the authorization by the museum to the agent or the authorization platform, and the second time is the authorization by the agent or the authorization platform to the third party. The process of this type of authorization mode is shown in Figure 2. The Louvre Museum in France and the Solomon R Guggenheim Museum in the United States are typical representatives of this authorization model.

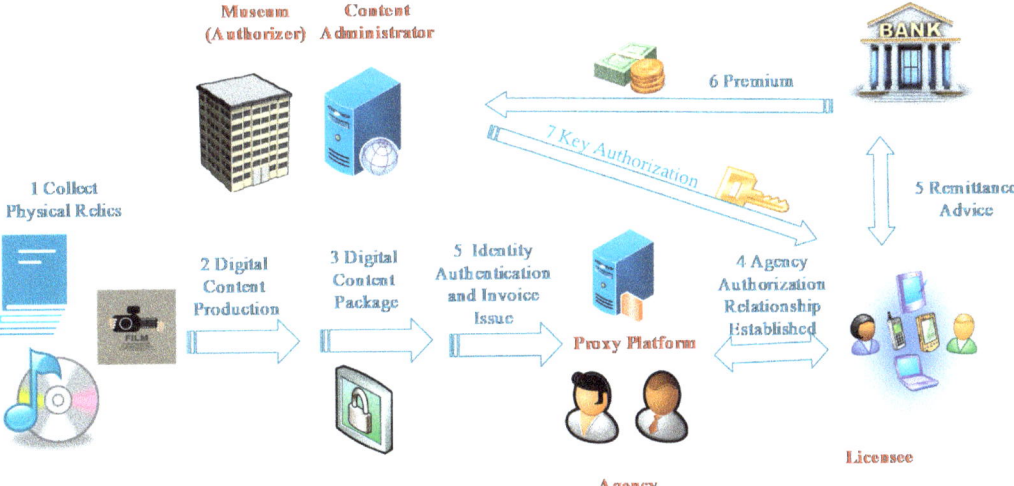

Figure 2. The digital direct authorization model of museum collections.

The entrusted authorization model means that the museum authorizes an agent to sign an authorization contract with the authorized person on behalf of the museum, a common museum proxy authorization model. In the proxy authorization model, agents as authorized intermediaries often have rich authorization management experience and mature customer groups, respond quickly to market demand, and have strong marketing capabilities, which can assist museums in rapidly opening up the authorization market, thereby promoting museums. The cultural and creative production industry has developed rapidly. However, agents, as market entities dominated by economic interests, tend to ignore the public welfare contained in cultural relics, significantly weakening the museum's ability to control the use of the digital collection by authorized third parties. In this process, third parties are based on market interests driving the development and utilization of authorized resources, so the cultural and economic risks faced by museums will increase accordingly.

The platform authorization model is similar to the entrusted authorization model, but there are differences in the scale of the authorizing party and the authorized party. Under the entrusted authorization model, it is usually one-to-one, that is, a museum entrusts a company to externally authorize, while under the platform authorization model, it is usually many-to-many, that is, multiple museums, middlemen, and authorized parties concentrated in a certain platform carry out authorization. The platform authorization model not only solves the problem of insufficient hardware facilities when most museums carry out the authorization of digital image resources of cultural relics, but also effectively reduces the transaction cost in the process of authorization of digital cultural relics. However, in the platform authorization model, the digital authorization of collections is mainly carried out in the network environment, which is likely to entail transaction risks, including intellectual property rights infringement.

3   Comprehensive authorization model for museum digital collections

The comprehensive authorization model is a composite authorization model, which is a diversified and differentiated authorization strategy made by the museum based on its actual situation. Possessing a certain brand awareness, a large number of collection images, high social recognition, and a variety of types of authorized objects are necessary conditions for the adoption of a comprehensive authorization model; therefore, it needs to be based on the museum's brand awareness, social influence, collection scale, and organization factors, such as staffing and the type of the subject matter of authorization, in making the relevant decision. The comprehensive authorization model combines the advantages of direct authorization and entrusted authorization and helps to optimize the authorization model of different subjects and maximize value creation. The disadvantage is that the complexity of the comprehensive authorization model increases the transaction cost of the authorization process, which will occupy more museum resources to a certain extent. The Metropolitan Museum of Art adopted a comprehensive authorization model when developing art authorization.

Blockchain is a kind of distributed data storage, which has the characteristics of point-to-point transmission, consensus mechanism, and encryption algorithm. For museums, blockchain technology has great value for the digitization of collections and artworks, especially cultural relics, specimens, and artworks. Blockchain has great potential in the confirmation of digital identities. This technology can generate an ID card based on an encryption algorithm for each institution or each person. It has the characteristics of decentralized data storage, decentralization, and traceability. Making clear value guarantees for each collection can also systematically protect the intellectual property rights of cultural relics and artworks so that the whole process of circulation can be followed. The production of digital content and the mechanism of cryptography comprise the foundation of digital property rights. In recent years, blockchain technology has been used to register and digitize collection-related information and cultural relic owner information, and then record these digital files on the blockchain. Because the blockchain has the characteristics of permanent storage and non-tampering, it can establish a one-to-one correspondence between collections, digital information (including photos, three-dimensional models, etc.), and owners, which can effectively solve cultural relic storage, ownership confirmation, and anti-theft, identification, loss prevention, and other issues.

This research is motivated by the following motivations:

(a) In the 20th century, international museums and governments, based on the mission of preservation and promotion of cultural policies to protect cultural resources, implemented digital plans for various museum collections, so that museums can share digital resources, which will not only help to promote social education, but also benefit the operation of museums.

(b) Under the guidance of the "activation and reproduction" thinking, this research uses a "digital authorization" model for museums to provide online users with information and increase financial resources to become a sustainable development of museum operations.

The main contributions of this work are as follows. This research proposes an authorization mechanism based on blockchain technology for protecting the museum's digital property rights. The signature and time stamp mechanism of cryptography is used to achieve a non-repudiation mechanism, and the smart contract achieves transparency, unforgeability, and traceability; this mechanism will thereby solving the above-mentioned problems faced by museum-digital rights management.

The rest of this article is organized as follows. The second section provides preliminary knowledge. The third section discusses the proposed methods for two kinds of authority mechanisms in the business model. The fourth section presents an analysis of the proposed scheme. The fifth section includes a discussion and comparison of the proposed scheme with related works. Finally, we present the conclusion and future works.

2. Preliminary

2.1. Smart Contract

A smart contract is a special agreement that is used when making a contract in the blockchain. It contains code functions and can interact with other contracts, guide decisions, store data, etc. The main force of smart contracts is to provide verification and execution of the conditions stipulated in the contract. Smart contracts allow credible transactions without the need for a third party. These transactions are traceable and irreversible. The concept of smart contracts was first proposed in 1994 by Nick Szabo [27,28], a computer scientist and cryptography expert. The purpose of smart contracts is to provide better security than traditional contract methods and to reduce other transaction costs associated with the contract.

2.2. ECDSA

In cryptography, the Elliptic Curve Digital Signature Algorithm (ECDSA) offers a variant of the Digital Signature Algorithm (DSA), which uses elliptic curve cryptography [29]. As with elliptic-curve cryptography in general, the bit size of the public key believed to be needed for ECDSA is about twice the size of the security level, in bits. For example, at a security level of 80 bits (meaning an attacker requires a maximum of about 2^{80} operations to find the private key), the size of an ECDSA public key would be 160 bits, whereas the size of a DSA public key is at least 1024 bits. On the other hand, the signature size is the same for both DSA and ECDSA: Approximately $4t$ bits, where t is the security level measured in bits; that is, about 320 bits for a security level of 80 bits.

The signature and verification process of ECDSA is as follows: Suppose Alice wants to send a message to Bob. Initially, both parties must reach a consensus on the curve parameters (CURVE, G, n). In addition to the field equation of the curve, the base point G on the curve and the multiplication order n of the base point G are also required. Alice also needs a private key, d_A and a public key, Q_A, where $Q_A = d_A G$. If the message Alice wants to send is m, Alice needs to choose a random value k between $[1, n-1]$: Calculate $z = h(m)$, $(x_1, y_1) = kG$, $r = x_1 \bmod n$, $s = k^{-1}(z + rd_A) \bmod n$, and send the ECDSA signature pair (r, s) together with the original message m to Bob. After receiving the signature pair (r, s) and the original message m, Bob will verify the correctness of the ECDSA signature. Bob first calculates $z' = h(m)$, $u_1 = z's^{-1} \bmod n$, $u_2 = rs^{-1} \bmod n$, $(x_1', y_1') = u_1 G + u_2 Q_A$, $r \stackrel{?}{=} x_1' \bmod n$, and if it passes the verification, then Bob confirms that the ECDSA signature and message m sent by Alice are correct.

2.3. Bilinear Pairings

The bilinear map was proposed by Boneh et al. in 2001 [30]. Later, Chen et al. applied this in the medical care field [31,32]. Let G_1 be a cyclic additive group generated by P, whose order is a prime q, and G_2 be a cyclic multiplicative group with the same order q. Let $e : G_1 * G_1 \rightarrow G_2$ be a map with the following properties:

(a) Bilinearity: $e(aP, bQ) = e(P, Q)^{ab}$, $P, Q \in G_1$, $a, b \in Z_q$.
(b) Non-degeneracy: There exists $P, Q \in G_1$ such that $e(P, Q) \neq 1$, in other words, the map does not send all pairs in $G_1 * G_1$ to the identity in G_2.
(c) Computability: There is an efficient algorithm to compute $e(P, Q)$, $P, Q \in G_1$.

2.4. Proxy Re-Encryption

In 1998, Blaze et al. [33] proposed atomic proxy cryptography for the first time, in which a semi-trusted proxy computes a function that converts ciphertexts for Alice into ciphertexts for Bob without seeing the underlying plaintext. In Elliptic Curve Based Proxy Re-Encryption, the authors combined elliptic curve, bilinear mapping, and proxy re-encryption and proposed the Elliptic Curve based proxy re-encryption. In their scheme, with setting up a large prime number and G, which is a point on elliptic curve E of order n, the proxy is entrusted with delegation key bG/a to change ciphertext from Alice to Bob via computing $(raGbG/a, rG^2 + P_m)$, where P_m is a point on the elliptic curve that embeds the message m in the elliptic curve equation f (i.e., $P_m = f(m)$).

Then we can calculate the message m by finding inverse as $f^{-1}(P_m)$. The proxy re-encryption is a natural application to secure the file system. The following scenarios are the Elliptic Curve based proxy re-encryption mechanism.

(a) System parameter establishment

Let E be an elliptic curve over a limited field F_q, where q is a large prime number, and G is a point on the elliptic curve E of order n. Let Z_n^* be a multiplicative group. Let the elliptic curve equation f denote the message embedding function, which maps the message m to a point P_m on E.

(b) Key generation

Alice randomly selects a positive integer $a \in Z_n^*$ as his/her private key and calculates aG as the public key. Bob randomly selects a positive integer $b \in Z_n^*$ as the private key and calculates bG as Bob's public key.

(c) Alice encrypts the plaintext m:
  1. P_m is the embedding message, which is calculated by $f(m)$: $P_m = f(m)$;
  2. generate an arbitrary number $r \in Z_n^*$ and output the ciphertext $(C_1, C_2) = (raG, rG^2 + P_m)$;
  3. send the ciphertext (C_1, C_2) to the proxy.

(d) Generation of the re-encryption key:
  1. Alice wants to authorize the information to Bob such that Bob can decrypt the ciphertext; Alice sends the proxy key $\pi_{A \rightarrow B} = bG/a$ to the proxy.
  2. The semi-honest agent proxy re-encrypts the ciphertext (C_1, C_2) into (C_1', C_2') and sends it to Bob.

(e) Re-encryption process:
  1. For the ciphertext $(C_1, C_2) = (raG, rG^2 + P_m)$, the proxy uses the re-encryption key to re-encrypt (C_1, C_2) into (C_1', C_2').
  2. (C_1', C_2')
     $= (raG\pi_{A \rightarrow B}, rG^2 + P_m)$
     $= (raGbG/a, rG^2 + P_m)$
     $= (rbG^2, rG^2 + P_m)$
  3. The proxy sends the converted ciphertext $(C_1', C_2') = (rbG^2, rG^2 + P_m)$ to Bob.

(f) Bob decrypts the ciphertext:
  1. Bob can decrypt the embedding message P_m with key b: $P_m = C_2' - b^{-1}C_1'$;
  2. then apply the inverse of the function f to get the original message m from P_m: $m = f^{-1}(P_m)$.

3. Method

3.1. System Architecture

Figure 3 is the system architecture diagram.

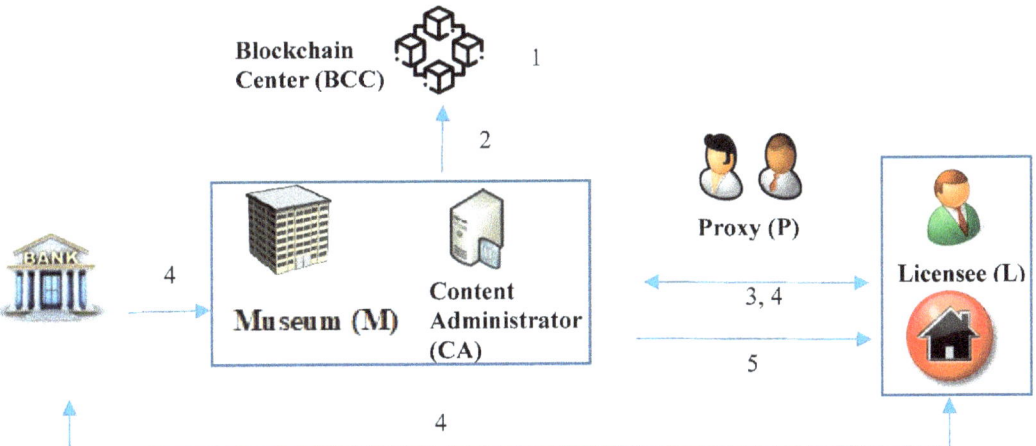

Figure 3. The system architecture.

In this study, we use the Elliptic Curve Digital Signature Algorithm (ECDSA), blockchain, and smart contracts to design a traceable authorization mechanism for the museum's digital content resource. There are six parties involved in this study: Museum (M), Content Administrator (CA), Licensee (L), Blockchain Center (BCC), Proxy (P), and Bank (B).

(a) Museum (M): The museum is the owner of the digital content. The museum collects the cultural relics and is responsible for the generation and management of the museum's digital content resource. The digital content resource is classified and protected by the museum.

(b) Content Administrator (CA): The CA is a cloud platform of the museum. It is responsible for reviewing the Licensee's request to determine 'allow or not' to access the digital content resource.

(c) Licensee (L): When citizens or institutions want to access the digital content resource of the museum, the Licensee should pay a premium to the museum.

(d) Blockchain Center (BCC): This center records the access information of the digital right resource for the Licensee. The BCC accepts the parties' registration and issues the identity certificate and public/private key pair to each party.

(e) Proxy (P): The proxy is an agency of the museum. After CA authenticates the Licensee's identity, P is responsible for actually cloud authorization for the Licensee to access the museum's digital content resource.

(f) Bank (B): Bank is authorized by a Licensee to pay a premium to the museum. We briefly illustrate the scenarios in the following steps.

- Step 1: Registration phase:

Museum, Licensee, Proxy, and Bank need to register with Blockchain Center; the Blockchain Center issues the identity certificate and public/private key pair to each party.

- Step 2: Digital content production phase:

The DCA classifies the museum's resources, encrypts these resources into a protected digital resource, and then stores it in the CA. The CA also uploads the detailed categories into the Blockchain center.

- Step 3: Authentication phase and issuing invoice phase:

After the Licensee proposes to access digital resource requests, the CA reviews the Licensee's qualifications and then issues the invoice.

- Step 4: Payment phase:

After payment, the Licensee requests the Bank to issue a certificate for the museum to authenticate this payment. The Content Administer then authenticates the Licensee's identity. The Content Administer performs one of the following cases.

Case 1: Generates the authorized key to the Licensee directly.

Case 2: Generates a proxy key to the Agency, and the Agency transfers it to the Licensee.

- Step 5: Digital content browsing phase:

After the Licensee receives the authorized key, the Licensee uses it to decrypt the protected digital content. The digital content can be read (or played) normally.

3.2. Smart Contract Initialization

In the proposed architecture, blockchain technology is applied. During the authentication and authorization process, some key information will be saved and verified through the blockchain. The key information in the blockchain is defined in the smart contract. The following is the blockchain smart contract structure for the proposed scheme (Scheme 1).

```
struct smart contract lminf/lainf/aminf{
    string lm/la/am id;
    string lm/la/am detail;
    string lm/la/am cert;
    string lm/la/am tsp;
}
struct smart contract mlinf/mainf/alinf {
    string ml/ma/al id;
    string ml/ma/al detail;
    string ml/ma/al tid;
    string ml/ma/al tsp;
}
```

```
struct smart contract lcinf/lpinf/pcinf{
    string lc/lp/pc id;
    string lc/lp/pc detail;
    string lc/lp/pc payment;
    string lc/lp/pc tsp;
}
struct smart contract clinf/cpinf/plinf {
    string cl/cp/pl id;
    string cl/cp/pl detail;
    string cl/cp/pl key;
    string cl/cp/pl tsp;
}
string keypairs;
string count;
```

Scheme 1. the blockchain smart contract initialization structure.

In the proposed smart contract, we have developed key information that will be stored in the blockchain. In the structure of the lm/la/am smart contract, we developed the field of id (identification), transaction detail, certificate, and timestamp. In the structure of the ml/ma/al smart contract, we developed the field of id, transaction detail, transaction id, and timestamp. In the structure of lc/lp/pc smart contract, we developed the field of id, transaction detail, payment information, and timestamp. In the structure of the cl/cp/pl smart contract, we developed the field of id, transaction detail, authentication key, and timestamp. In the initialization phase, the blockchain center also issues the public and private key pairs for all roles.

3.3. Registration Phase

The Licensee (L), Content Administrator (CA), and Proxy (P) should register with the Blockchain Center (BCC) and obtain a relative public/private key pair. The Licensee (L) and Proxy (P) also get a digital certificate of identity from the Blockchain Center via a secure channel. The system role X can represent the Licensee (L), Content Administrator (CA), and Proxy (P). Figure 4 shows the flowchart of the registration phase.

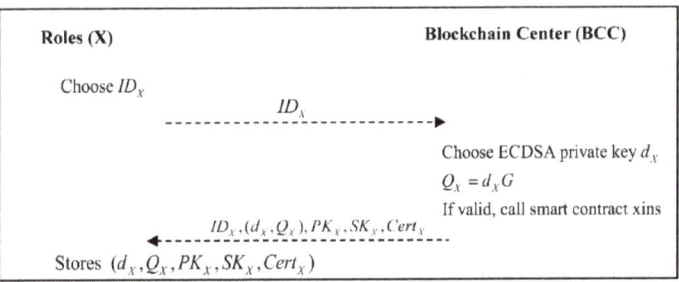

Figure 4. Each role of the system registers with the Blockchain Center.

- Step 1: Role X generates an identity ID_X, and sends it to the Blockchain Center.
- Step 2: The Blockchain center generates an ECDSA private key d_X based on the role X, calculates:

$$Q_X = d_X G. \tag{1}$$

If the identity of the registered role is verified, the smart contract Xins will be triggered, and the content is presented as follows (Scheme 2):

```
function insert x smart contract xins (
string x_id, string x_detail) {
    count ++;
    x[count].id = id;
    x[count].detail = detail;
}
string x_keypairs;
```

Scheme 2. The smart contract Xins.

Then the blockchain center will transmit $ID_X, (d_X, Q_X), PK_X, SK_X, Cert_X$ to role X.

- Step 3: The role X stores $(d_X, Q_X, PK_X, SK_X, Cert_X)$.

3.4. Digital Content Production Phase

The museum collects many precious cultural relics. The digital content production process of valuable cultural relics involves a specific process. In general, experts and scholars classify (such as biological classification, antiquities classification, etc.), grade (grade of antiquities is divided into general, important, national treasures, etc.), and clarify the importance (such as rare or era significance or endangered species, etc.), and then different competent authorities proceed with various kinds of appointments. Finally, it is handed over to professional and technical personnel to produce digital content through photography and 3D surroundings.

In this phase, we will focus on illustrating the protection technology of digital content. Figure 3 shows the production flowchart of protected digital content. To enhance performance, we use the digital envelope for implementation. That is, the Content Administrator (CA) uses the symmetry key to encrypt the digital content, and then uses the ElGamal-based system of the public-key system to protect the symmetry key. Figure 5 shows the flowchart of the digital content production phase.

- Step 1: Content Administrator (CA) collects cultural relics in a systematic and planned way according to the categories of different collections. CA also uses information technology to convert the collected media data into a form that can be stored, processed, and edited.
- Step 2: CA encrypts these encoded multimedia data with KeyID and Seed, organizes and categorizes each digitized archive resource, and records the data description of

the archive itself, as an annotation explanation for the archive itself and various media materials, as well as an indexing tool for users to inquire.
- Step 3: Through the overall planning of the collection environment, a suitable information system can be constructed, and the functions of digital data preservation and management can be achieved through the operation of the system. When a Licensee wants to access these multimedia materials, it must first obtain legal authorization from the Content Administrator (CA).
- Step 4: The CA will provide the Licensee with an authorization key; the Licensee can use the authorization key to unlock the information provided by the CA and get a decryption key, which can be used to obtain the plaintext of multimedia messages. The details will be introduced in the following phase.

Figure 5. Digital content production phase.

3.5. Authentication and Issuing Invoice Phase

3.5.1. Case 1: Direct Authorization

After reviewing the Licensee's identity, the Content Administrator generates a transaction ID and invoice to the Licensee. We present the flowchart of the authentication and issuing an invoice phase for direct authorization in Figure 6.

- Step 1: The Licensee generates a random value k_{L-M}, calculates:

$$z_{L-M} = h(ID_L, M_{L-M}, Cert_L, TS_{L-M}, ID_{BC}), \quad (2)$$

$$(x_{L-M}, y_{L-M}) = k_{L-M}G, \quad (3)$$

$$r_{L-M} = x_{L-M} \bmod n, \quad (4)$$

$$s_{L-M} = k_{L-M}^{-1}(z_{L-M} + r_{L-M}d_L) \bmod n, \quad (5)$$

$$Enc_{L-M} = E_{PK_M}(ID_L, M_{L-M}, Cert_L, TS_{L-M}, ID_{BC}), \quad (6)$$

and sends $ID_L, Enc_{L-M}, (r_{L-M}, s_{L-M})$ to the content administrator.

The ID_L is encrypted to check integrity. The second ID_L is to show the Licensee's identity to the content administrator.

- Step 2: The Content Administrator first calculates:

$$(ID_L, M_{L-M}, Cert_L, TS_{L-M}, ID_{BC}) = D_{SK_M}(Enc_{L-M}), \quad (7)$$

uses

$$TS_{NOW} - TS_{L-M} \leq \Delta T \quad (8)$$

to confirm whether the timestamp is valid, verifies $Cert_L$ with PK_L, verifies the correctness of the ECDSA signature, then calculates:

$$z_{L-M}' = h(ID_L, M_{L-M}, Cert_L, TS_{L-M}, ID_{BC}), \quad (9)$$

$$u_{L-M1} = z_{L-M}'s_{L-M}^{-1} \bmod n, \quad (10)$$

$$u_{L-M2} = r_{L-M}s_{L-M}^{-1} \bmod n, \quad (11)$$

$$(x_{L-M}', y_{L-M}') = u_{L-M1}G + u_{L-M2}Q_L, \quad (12)$$

$$x_{L-M}' \stackrel{?}{=} r_{L-M} \bmod n. \quad (13)$$

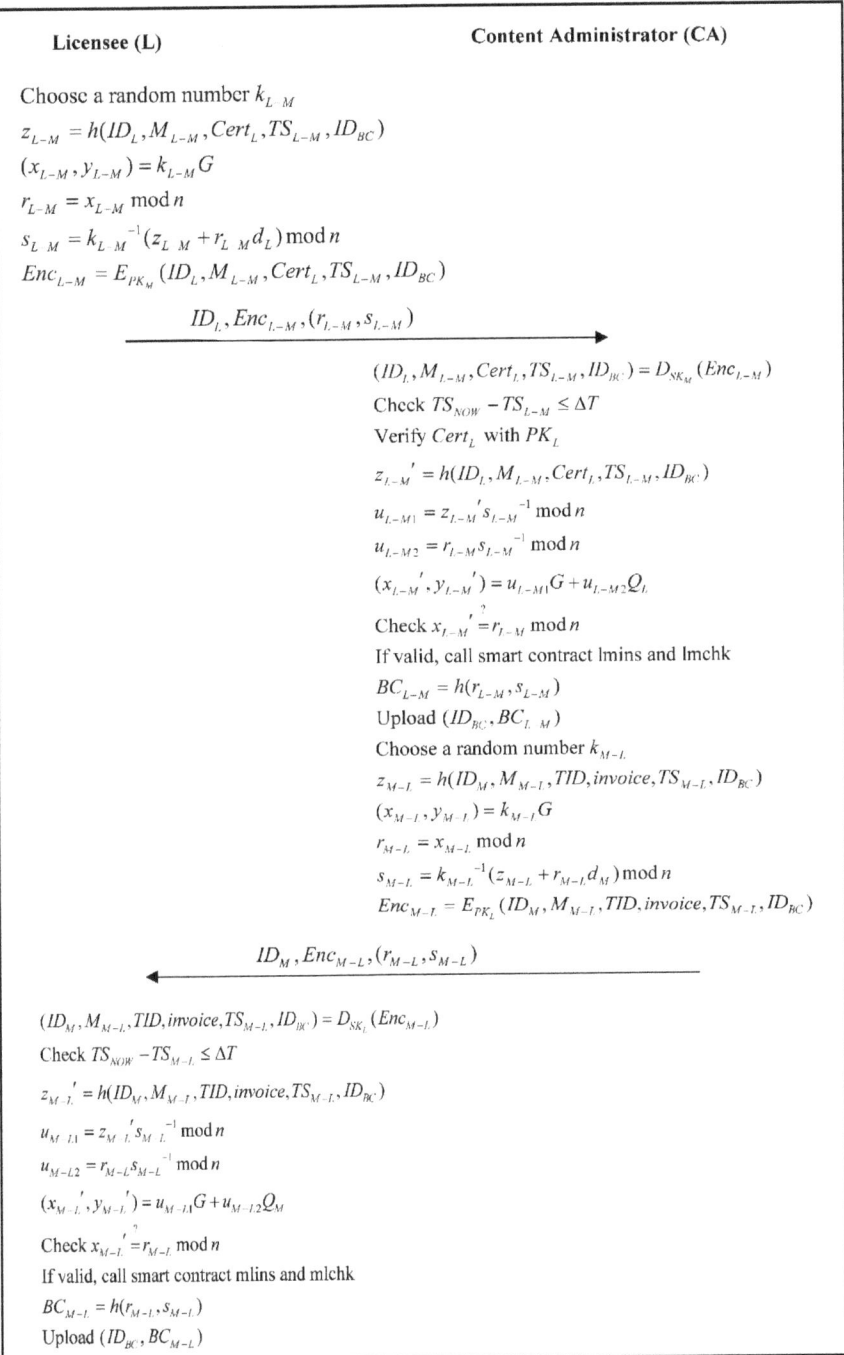

Figure 6. Authentication and issuing invoice phase (direct authorization).

If the verification is passed, CA will get the relevant content request information and trigger the smart contracts lmins and lmchk. The content is as follows (Scheme 3):

```
function insert smart contract lmins(
string lm_id, string lm_detail,
string lm_cert, string lm_tsp) {
    count ++;
    lm[count].id = id;
    lm[count].detail = detail;
    lm[count].cert = cert;
    lm[count].tsp = tsp;
}
sign string l_key (lm_id, lm_detail,
lm_cert, lm_tsp);
```

```
verify string l_key (lm_id, lm_detail,
lm_cert, lm_tsp);
function check smart contract lmchk(
string lm_id, string lm_detail,
string lm_cert, string lm_tsp) {
    return lm_id.exist;
    return lm_detail.exist;
    return lm_cert.exist;
    return lm_tsp.exist;
}
```

Scheme 3. The smart contracts lmins and lmchk.

The CA calculates:
$$BC_{L-M} = h(r_{L-M}, s_{L-M}), \tag{14}$$

(ID_{BC}, BC_{L-M}) will also be uploaded to the blockchain center. Then the CA generates a random value k_{M-L} and calculates:

$$z_{M-L} = h(ID_M, M_{M-L}, TID, invoice, TS_{M-L}, ID_{BC}), \tag{15}$$

$$(x_{M-L}, y_{M-L}) = k_{M-L}G, \tag{16}$$

$$r_{M-L} = x_{M-L} \bmod n, \tag{17}$$

$$s_{M-L} = k_{M-L}^{-1}(z_{M-L} + r_{M-L}d_M) \bmod n, \tag{18}$$

$$Enc_{M-L} = E_{PK_L}(ID_M, M_{M-L}, TID, invoice, TS_{M-L}, ID_{BC}), \tag{19}$$

and sends $ID_M, Enc_{M-L}, (r_{M-L}, s_{M-L})$ to the Licensee.

- Step 3: The Licensee first calculates:

$$(ID_M, M_{M-L}, TID, invoice, TS_{M-L}, ID_{BC}) = D_{SK_L}(Enc_{M-L}), \tag{20}$$

uses
$$TS_{NOW} - TS_{M-L} \leq \Delta T \tag{21}$$

to confirm whether the timestamp is valid, verifies the correctness of the ECDSA signature, then calculates:

$$z_{M-L}' = h(IID_M, M_{M-L}, TID, invoice, TS_{M-L}, ID_{BC}), \tag{22}$$

$$u_{M-L1} = z_{M-L}' s_{M-L}^{-1} \bmod n, \tag{23}$$

$$u_{M-L2} = r_{M-L} s_{M-L}^{-1} \bmod n, \tag{24}$$

$$(x_{M-L}', y_{M-L}') = u_{M-L1}G + u_{M-L2}Q_M, \tag{25}$$

$$x_{M-L}' \stackrel{?}{=} r_{M-L} \bmod n. \tag{26}$$

If the verification is passed, the content request information is confirmed by CA, and the smart contracts mlins and mlchk will be sent. The content is as follows (Scheme 4):

```
function insert smart contract mlins(
string ml_id, string ml_detail,
string ml_tid, string ml_tsp) {
    count ++;
    ml[count].id = id;
    ml[count].detail = detail;
    ml[count].tid = tid;
    ml[count].tsp = tsp;
}
sign string m_key (ml_id, ml_detail,
ml_tid, ml_tsp);
```

```
verify string m_key (ml_id, ml_detail,
ml_tid, ml_tsp);
function check smart contract mlchk(
string ml_id, string ml_detail,
string ml_tid, string ml_tsp) {
    return ml_id.exist;
    return ml_detail.exist;
    return ml_tid.exist;
    return ml_tsp.exist;
}
```

Scheme 4. The smart contracts mlins and mlchk.

The Licensee calculates:

$$BC_{M-L} = h(r_{M-L}, s_{M-L}), \qquad (27)$$

(ID_{BC}, BC_{M-L}) will also be uploaded to the blockchain center.

3.5.2. Case 2: Proxy Authorization

When the Licensee submits an application request to the Proxy, the Proxy transfers it to the CA for verification. After reviewing the Licensee's identity, the CA generates a transaction ID and invoice to the Licensee. We present the flowchart of the authentication and issuing invoice phase (L to P) in Figure 7, the flowchart of the authentication and issuing invoice phase (P to CA) in Figure 8, the flowchart of the authentication and issuing invoice phase (CA to P) in Figure 9, and the flowchart of the authentication and issuing invoice phase (P to L) in Figure 10.

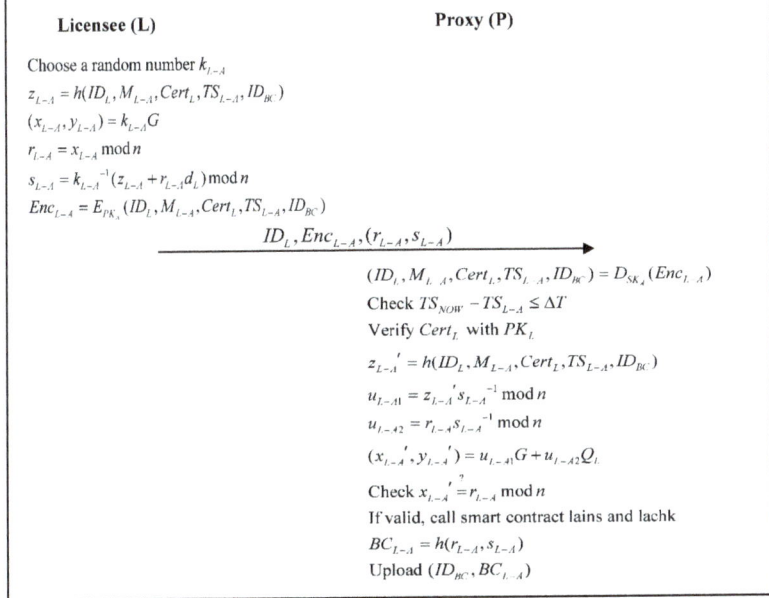

Figure 7. Authentication and issuing invoice phase (L to P).

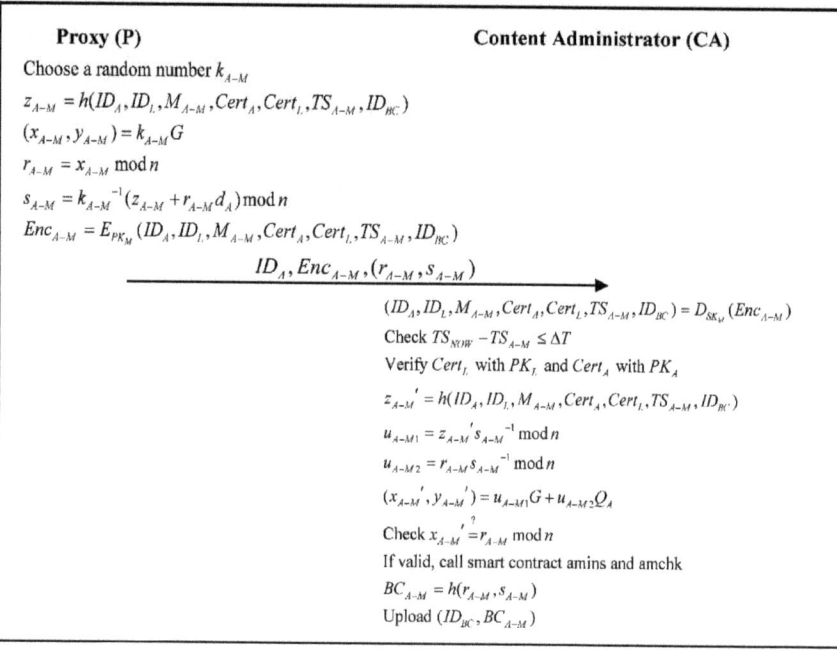

Figure 8. Authentication and issuing invoice phase (P to CA).

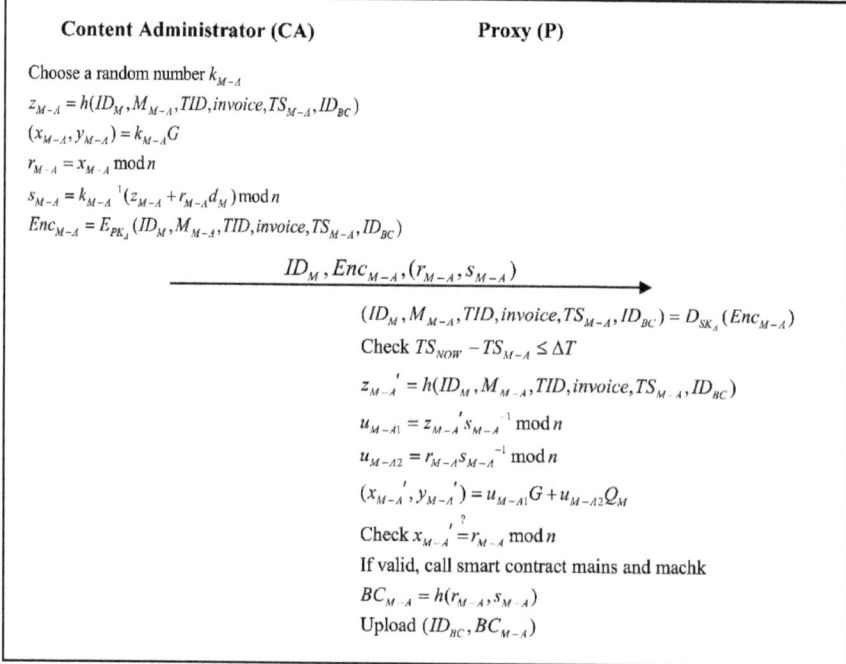

Figure 9. Authentication and issuing invoice phase (CA to P).

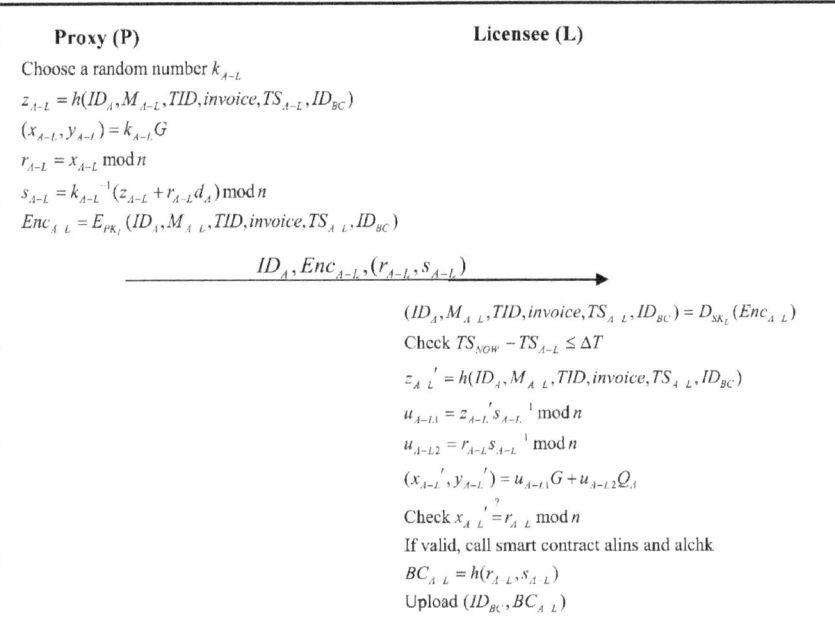

Figure 10. Authentication and issuing invoice phase (P to L).

- Step 1: The Licensee generates a random value k_{L-A}, calculates:

$$z_{L-A} = h(ID_L, M_{L-A}, Cert_L, TS_{L-A}, ID_{BC}), \qquad (28)$$

$$(x_{L-A}, y_{L-A}) = k_{L-A}G, \qquad (29)$$

$$r_{L-A} = x_{L-A} \bmod n, \qquad (30)$$

$$s_{L-A} = k_{L-A}^{-1}(z_{L-A} + r_{L-A}d_L) \bmod n, \qquad (31)$$

$$Enc_{L-A} = E_{PK_A}(ID_L, M_{L-A}, Cert_L, TS_{L-A}, ID_{BC}), \qquad (32)$$

and sends $ID_L, Enc_{L-A}, (r_{L-A}, s_{L-A})$ to the proxy.

- Step 2: The proxy first calculates:

$$(ID_L, M_{L-A}, Cert_L, TS_{L-A}, ID_{BC}) = D_{SK_A}(Enc_{L-A}), \qquad (33)$$

uses

$$TS_{NOW} - TS_{L-A} \leq \Delta T \qquad (34)$$

to confirm whether the timestamp is valid, verifies $Cert_L$ with PK_L, verifies the correctness of the ECDSA signature, and then calculates:

$$z_{L-A}' = h(ID_L, M_{L-A}, Cert_L, TS_{L-A}, ID_{BC}), \qquad (35)$$

$$u_{L-A1} = z_{L-A}' s_{L-A}^{-1} \bmod n, \qquad (36)$$

$$u_{L-A2} = r_{L-A} s_{L-A}^{-1} \bmod n, \qquad (37)$$

$$(x_{L-A}', y_{L-A}') = u_{L-A1}G + u_{L-A2}Q_L, \qquad (38)$$

$$x_{L-A}' \stackrel{?}{=} r_{L-A} \bmod n. \qquad (39)$$

If the verification is passed, the proxy will get the relevant content request information and trigger the smart contracts lains and lachk. The content is as follows (Scheme 5):

```
function insert smart contract lains(
string la_id, string la_detail,
string la_cert, string la_tsp) {
    count ++;
    la[count].id = id;
    la[count].detail = detail;
    la[count].cert = cert;
    la[count].tsp = tsp;
}
sign string l_key (la_id, la_detail,
la_cert, la_tsp);
```

```
verify string l_key (la_id, la_detail,
la_cert, la_tsp);
function check smart contract lachk(
string la_id, string la_detail,
string la_cert, string la_tsp) {
    return la_id.exist;
    return la_detail.exist;
    return la_cert.exist;
    return la_tsp.exist;
}
```

Scheme 5. The smart contracts lains and lachk.

The proxy calculates:

$$BC_{L-A} = h(r_{L-A}, s_{L-A}), \qquad (40)$$

(ID_{BC}, BC_{L-A}) will also be uploaded to the blockchain center.

- Step 3: The proxy generates a random value k_{A-M} and calculates:

$$z_{A-M} = h(ID_A, ID_L, M_{A-M}, Cert_A, Cert_L, TS_{A-M}, ID_{BC}), \qquad (41)$$

$$(x_{A-M}, y_{A-M}) = k_{A-M}G, \qquad (42)$$

$$r_{A-M} = x_{A-M} \bmod n, \qquad (43)$$

$$s_{A-M} = k_{A-M}^{-1}(z_{A-M} + r_{A-M}d_A) \bmod n, \qquad (44)$$

$$Enc_{A-M} = E_{PK_M}(ID_A, ID_L, M_{A-M}, Cert_A, Cert_L, TS_{A-M}, ID_{BC}), \qquad (45)$$

and sends $ID_A, Enc_{A-M}, (r_{A-M}, s_{A-M})$ to the CA.

- Step 4: The CA first calculates:

$$(ID_A, ID_L, M_{A-M}, Cert_A, Cert_L, TS_{A-M}, ID_{BC}) = D_{SK_M}(Enc_{A-M}), \qquad (46)$$

uses

$$TS_{NOW} - TS_{A-M} \leq \Delta T \qquad (47)$$

to confirm whether the timestamp is valid, verifies $Cert_L$ with PK_L and $Cert_A$ with PK_A, verifies the correctness of the ECDSA signature, and then calculates:

$$z_{A-M}' = h(ID_A, ID_L, M_{A-M}, Cert_A, Cert_L, TS_{A-M}, ID_{BC}), \qquad (48)$$

$$u_{A-M1} = z_{A-M}' s_{A-M}^{-1} \bmod n, \qquad (49)$$

$$u_{A-M2} = r_{A-M} s_{A-M}^{-1} \bmod n, \qquad (50)$$

$$(x_{A-M}', y_{A-M}') = u_{A-M1}G + u_{A-M2}Q_A, \qquad (51)$$

$$x_{A-M}' \stackrel{?}{=} r_{A-M} \bmod n. \qquad (52)$$

If the verification is passed, the CA will get the relevant content request information and trigger the smart contracts amins and amchk. The content is as follows (Scheme 6):

```
function insert smart contract amins(
string am_id, string am_detail,
string am_cert, string am_tsp) {
    count ++;
     am[count].id = id;
     am[count].detail = detail;
     am[count].cert = cert;
     am[count].tsp = tsp;
}
sign string a_key (am_id, am_detail,
am_cert, am_tsp);
```

```
verify string a_key (am_id, am_detail,
am_cert, am_tsp);
function check smart contract amchk(
string am_id, string am_detail,
string am_cert, string am_tsp) {
     return am_id.exist;
     return am_detail.exist;
     return am_cert.exist;
     return am_tsp.exist;
}
```

Scheme 6. The smart contracts amins and amchk.

The CA calculates:
$$BC_{A-M} = h(r_{A-M}, s_{A-M}), \tag{53}$$

(ID_{BC}, BC_{A-M}) will also be uploaded to the blockchain center.

- Step 5: The CA generates a random value k_{M-A} and calculates:

$$z_{M-A} = h(ID_M, M_{M-A}, TID, invoice, TS_{M-A}, ID_{BC}), \tag{54}$$

$$(x_{M-A}, y_{M-A}) = k_{M-A}G, \tag{55}$$

$$r_{M-A} = x_{M-A} \bmod n, \tag{56}$$

$$s_{M-A} = k_{M-A}^{-1}(z_{M-A} + r_{M-A}d_M) \bmod n, \tag{57}$$

$$Enc_{M-A} = E_{PK_A}(ID_M, M_{M-A}, TID, invoice, TS_{M-A}, ID_{BC}), \tag{58}$$

and sends $ID_M, Enc_{M-A}, (r_{M-A}, s_{M-A})$ to the proxy.

- Step 6: The proxy first calculates:

$$(ID_M, M_{M-A}, TID, invoice, TS_{M-A}, ID_{BC}) = D_{SK_A}(Enc_{M-A}), \tag{59}$$

uses
$$TS_{NOW} - TS_{M-A} \leq \Delta T \tag{60}$$

to confirm whether the timestamp is valid, verifies the correctness of the ECDSA signature, and then calculates:

$$z_{M-A}' = h(ID_M, M_{M-A}, TID, invoice, TS_{M-A}, ID_{BC}), \tag{61}$$

$$u_{M-A1} = z_{M-A}' s_{M-A}^{-1} \bmod n, \tag{62}$$

$$u_{M-A2} = r_{M-A} s_{M-A}^{-1} \bmod n, \tag{63}$$

$$(x_{M-A}', y_{M-A}') = u_{M-A1}G + u_{M-A2}Q_M, \tag{64}$$

$$x_{M-A}' \stackrel{?}{=} r_{M-A} \bmod n. \tag{65}$$

If the verification is passed, the content request information is confirmed by the proxy, and the smart contracts mains and machk will be sent. The content is as follows (Scheme 7):

```
function insert smart contract mains(
string ma_id, string ma_detail,
string ma_tid, string ma_tsp) {
    count ++;
    ma[count].id = id;
    ma[count].detail = detail;
    ma[count].tid = tid;
    ma[count].tsp = tsp;
}
sign string m_key (ma_id, ma_detail,
ma_tid, ma_tsp);
```

```
verify string m_key (ma_id, ma_detail,
ma_tid, ma_tsp);
function check smart contract machk(
string ma_id, string ma_detail,
string ma_tid, string ma_tsp) {
    return ma_id.exist;
    return ma_detail.exist;
    return ma_tid.exist;
    return ma_tsp.exist;
}
```

Scheme 7. The smart contracts mains and machk.

The proxy calculates:

$$BC_{M-A} = h(r_{M-A}, s_{M-A}), \tag{66}$$

(ID_{BC}, BC_{M-A}) will also be uploaded to the blockchain center.

- Step 7: The proxy generates a random value k_{A-L} and calculates:

$$z_{A-L} = h(ID_A, M_{A-L}, TID, invoice, TS_{A-L}, ID_{BC}), \tag{67}$$

$$(x_{A-L}, y_{A-L}) = k_{A-L}G, \tag{68}$$

$$r_{A-L} = x_{A-L} \bmod n, \tag{69}$$

$$s_{A-L} = k_{A-L}^{-1}(z_{A-L} + r_{A-L}d_A) \bmod n, \tag{70}$$

$$Enc_{A-L} = E_{PK_L}(ID_A, M_{A-L}, TID, invoice, TS_{A-L}, ID_{BC}), \tag{71}$$

and sends $ID_A, Enc_{A-L}, (r_{A-L}, s_{A-L})$ to the Licensee.

- Step 8: The Licensee first calculates:

$$(ID_A, M_{A-L}, TID, invoice, TS_{A-L}, ID_{BC}) = D_{SK_L}(Enc_{A-L}), \tag{72}$$

uses

$$TS_{NOW} - TS_{A-L} \leq \Delta T \tag{73}$$

to confirm whether the timestamp is valid, verifies the correctness of the ECDSA signature, and then calculates:

$$z_{A-L}' = h(ID_A, M_{A-L}, TID, invoice, TS_{A-L}, ID_{BC}), \tag{74}$$

$$u_{A-L1} = z_{A-L}' s_{A-L}^{-1} \bmod n, \tag{75}$$

$$u_{A-L2} = r_{A-L} s_{A-L}^{-1} \bmod n, \tag{76}$$

$$(x_{A-L}', y_{A-L}') = u_{A-L1}G + u_{A-L2}Q_A, \tag{77}$$

$$x_{A-L}' \stackrel{?}{=} r_{A-L} \bmod n. \tag{78}$$

If the verification is passed, the content request information is confirmed by the CA, and the smart contracts alins and alchk will be sent. The content is as follows (Scheme 8):

```
function insert smart contract alins(
string al_id, string al_detail,
string al_tid, string al_tsp) {
    count ++;
    al[count].id = id;
    al[count].detail = detail;
    al[count].tid = tid;
    al[count].tsp = tsp;
}
sign string a_key (al_id, al_detail,
al_tid, al_tsp);
```

```
verify string a_key (al_id, al_detail,
al_tid, al_tsp);
function check smart contract alchk(
string al_id, string al_detail,
string al_tid, string al_tsp) {
    return al_id.exist;
    return al_detail.exist;
    return al_tid.exist;
    return al_tsp.exist;
}
```

Scheme 8. The smart contracts alins and alchk.

The Licensee calculates:

$$BC_{A-L} = h(r_{A-L}, s_{A-L}), \tag{79}$$

(ID_{BC}, BC_{A-L}) will also be uploaded to the blockchain center.

3.6. Payment Verification and Browsing Phase

3.6.1. Case 1: Direct Authorization

After the Licensee is paid, the bank must sign and issue the payment certificate to verify. The CA then authenticates the Licensee's identity and bank payment certificate. After that, the CA generates the authorized key, making time-sensitive tokens. After authorization, the Licensee's application (reader or player) can use the authorized key to automatically decrypt the symmetry key. The APP can browse digital content normally. We present the flowchart of the payment verification and browsing phase for direct authorization in Figure 11.

- Step 1: The Licensee generates a random value k_{L-C}, calculates:

$$z_{L-C} = h(ID_L, M_{L-C}, Cert_L, TID, Cert_{pay}, TS_{L-C}, ID_{BC}), \tag{80}$$

$$(x_{L-C}, y_{L-C}) = k_{L-C}G, \tag{81}$$

$$r_{L-C} = x_{L-C} \bmod n, \tag{82}$$

$$s_{L-C} = k_{L-C}^{-1}(z_{L-C} + r_{L-C}d_L) \bmod n, \tag{83}$$

$$Enc_{L-C} = E_{PK_C}(ID_L, M_{L-C}, Cert_L, TID, Cert_{pay}, TS_{L-C}, ID_{BC}), \tag{84}$$

and sends $ID_L, Enc_{L-C}, (r_{L-C}, s_{L-C})$ to the content administrator.

- Step 2: The CA first calculates:

$$(ID_L, M_{L-C}, Cert_L, TID, Cert_{pay}, TS_{L-C}, ID_{BC}) = D_{SK_C}(Enc_{L-C}), \tag{85}$$

uses

$$TS_{NOW} - TS_{L-C} \leq \Delta T \tag{86}$$

to confirm whether the timestamp is valid, it verifies $Cert_L$ with PK_L and $Cert_{pay}$ with PK_{BANK}, verifies the correctness of the ECDSA signature, and then calculates:

$$z_{L-C}' = h(ID_L, M_{L-C}, Cert_L, TID, Cert_{pay}, TS_{L-C}, ID_{BC}), \tag{87}$$

$$u_{L-C1} = z_{L-C}' s_{L-C}^{-1} \bmod n, \tag{88}$$

$$u_{L-C2} = r_{L-C}s_{L-C}^{-1} \bmod n, \tag{89}$$

$$(x_{L-C}', y_{L-C}') = u_{L-C1}G + u_{L-C2}Q_L, \tag{90}$$

$$x_{L-C}' \stackrel{?}{=} r_{L-C} \bmod n. \tag{91}$$

```
┌─────────────────────────────────────────────────────────────────────┐
│   Licensee (L)                          Content Administrator (CA)  │
│                                                                     │
│ Choose a random number k_{L-C}                                      │
│ z_{L-C} = h(ID_L, M_{L-C}, Cert_L, TID, Cert_{pay}, TS_{L-C}, ID_{BC}) │
│ (x_{L-C}, y_{L-C}) = k_{L-C}G                                       │
│ r_{L-C} = x_{L-C} mod n                                             │
│ s_{L-C} = k_{L-C}^{-1}(z_{L-C} + r_{L-C}d_L) mod n                  │
│ Enc_{L-C} = E_{PK_C}(ID_L, M_{L-C}, Cert_L, TID, Cert_{pay}, TS_{L-C}, ID_{BC}) │
│                                                                     │
│       ID_L, Enc_{L-C}, (r_{L-C}, s_{L-C})                           │
│       ─────────────────────────────────────►                        │
│                                                                     │
│              (ID_L, M_{L-C}, Cert_L, TID, Cert_{pay}, TS_{L-C}, ID_{BC}) = D_{SK_C}(Enc_{L-C}) │
│              Check TS_{NOW} − TS_{L-C} ≤ ΔT                         │
│              Verify Cert_L with PK_I and Cert_{pay} with PK_{BANK}  │
│              z_{L-C}' = h(ID_L, M_{L-C}, Cert_L, TID, Cert_{pay}, TS_{L-C}, ID_{BC}) │
│              u_{L-C1} = z_{L-C}' s_{L-C}^{-1} mod n                 │
│              u_{L-C2} = r_{L-C} s_{L-C}^{-1} mod n                  │
│              (x_{L-C}', y_{L-C}') = u_{L-C1}G + u_{L-C2}Q_L         │
│              Check x_{L-C}' =? r_{L-C} mod n                        │
│              If valid, call smart contract lcins and lcchk          │
│              BC_{L-C} = h(r_{L-C}, s_{L-C})                         │
│              Upload (ID_{BC}, BC_{L-C})                             │
│              Choose a random number k_{C-L}                         │
│              P_m = f(ID_{DC}, Vtime)                                │
│              C_1 = Z^{rb}G                                          │
│              C_2 = Z'G + P_m                                        │
│              z_{C-L} = h(ID_C, M_{C-L}, (C_1, C_2), TS_{C-L}, ID_{BC}) │
│              (x_{C-L}, y_{C-L}) = k_{C-L}G                          │
│              r_{C-L} = x_{C-L} mod n                                │
│              s_{C-L} = k_{C-L}^{-1}(z_{C-L} + r_{C-L}d_C) mod n     │
│              Enc_{C-L} = E_{PK_L}(ID_C, M_{C-L}, (C_1, C_2), TS_{C-L}, ID_{BC}) │
│                                                                     │
│              ID_C, Enc_{C-L}, (r_{C-L}, s_{C-L})                    │
│       ◄─────────────────────────────────────                        │
│                                                                     │
│ (ID_C, M_{C-L}, (C_1,C_2), TS_{C-L}, ID_{BC}) = D_{SK_L}(Enc_{C-L}) │
│ Check TS_{NOW} − TS_{C-L} ≤ ΔT                                      │
│ z_{C-L}' = h(ID_C, M_{C-L}, (C_1,C_2), TS_{C-L}, ID_{BC})           │
│ u_{C-L1} = z_{C-L}' s_{C-L}^{-1} mod n                              │
│ u_{C-L2} = r_{C-L} s_{C-L}^{-1} mod n                               │
│ (x_{C-L}', y_{C-L}') = u_{C-L1}G + u_{C-L2}Q_C                      │
│ Check x_{C-L}' =? r_{C-L} mod n                                     │
│ If valid, call smart contract clins and clchk                       │
│ BC_{C-L} = h(r_{C-L}, s_{C-L})                                      │
│ Upload (ID_{BC}, BC_{C-L})                                          │
│ Calculate P_m = C_2 − C_1^b through APP                             │
└─────────────────────────────────────────────────────────────────────┘
```

Figure 11. Payment verification and browsing phase (direct authorization).

If the verification is passed, the content administrator will get the relevant payment information and trigger the smart contracts lcins and lcchk. The content is as follows (Scheme 9):

```
function insert smart contract lcins(
string lc_id, string lc_detail,
string lc_payment, string lc_tsp) {
    count ++;
    lc[count].id = id;
    lc[count].detail = detail;
    lc[count].payment = payment;
    lc[count].tsp = tsp;
}
sign string l_key (lc_id, lc_detail,
lc_payment, lc_tsp);
```

```
verify string l_key (lc_id, lc_detail,
lc_payment, lc_tsp);
function check smart contract lcchk(
string lc_id, string lc_detail,
string lc_payment, string lc_tsp) {
    return lc_id.exist;
    return lc_detail.exist;
    return lc_payment.exist;
    return lc_tsp.exist;
}
```

Scheme 9. The smart contracts lcins and lcchk.

The content administrator calculates:

$$BC_{L-C} = h(r_{L-C}, s_{L-C}), \tag{92}$$

(ID_{BC}, BC_{L-C}) will also be uploaded to the blockchain center. Then the content administrator generates a random value k_{C-L} and calculates:

$$P_m = f(ID_{DC}, Vtime), \tag{93}$$

$$C_1 = Z^{rb}G, \tag{94}$$

$$C_2 = Z^r G + P_m, \tag{95}$$

$$z_{C-L} = h(ID_C, M_{C-L}, (C_1, C_2), TS_{C-L}, ID_{BC}), \tag{96}$$

$$(x_{C-L}, y_{C-L}) = k_{C-L}G, \tag{97}$$

$$r_{C-L} = x_{C-L} \bmod n, \tag{98}$$

$$s_{C-L} = k_{C-L}^{-1}(z_{C-L} + r_{C-L}d_C) \bmod n, \tag{99}$$

$$Enc_{C-L} = E_{PK_L}(ID_C, M_{C-L}, (C_1, C_2), TS_{C-L}, ID_{BC}), \tag{100}$$

and sends $ID_C, Enc_{C-L}, (r_{C-L}, s_{C-L})$ to the Licensee.

- Step 3: The Licensee first calculates:

$$(ID_C, M_{C-L}, (C_1, C_2), TS_{C-L}, ID_{BC}) = D_{SK_L}(Enc_{C-L}), \tag{101}$$

uses

$$TS_{NOW} - TS_{C-L} \le \Delta T \tag{102}$$

to confirm whether the timestamp is valid, verifies the correctness of the ECDSA signature, and then calculates:

$$z_{C-L}' = h(ID_C, M_{C-L}, (C_1, C_2), TS_{C-L}, ID_{BC}), \tag{103}$$

$$u_{C-L1} = z_{C-L}' s_{C-L}^{-1} \bmod n, \tag{104}$$

$$u_{C-L2} = r_{C-L} s_{C-L}^{-1} \bmod n, \tag{105}$$

$$(x_{C-L}', y_{C-L}') = u_{C-L1}G + u_{C-L2}Q_C, \tag{106}$$

$$x_{C-L}' \stackrel{?}{=} r_{C-L} \bmod n. \tag{107}$$

If the verification is passed, the payment information is confirmed by the content administrator, and the smart contracts clins and clchk will be sent. The content is as follows (Scheme 10):

```
function insert smart contract clins(
string cl_id, string cl_detail,
string cl_key, string cl_tsp) {
    count ++;
    cl[count].id = id;
    cl[count].detail = detail;
    cl[count].key = key;
    cl[count].tsp = tsp;
}
sign string c_key (cl_id, cl_detail,
cl_key, cl_tsp);
```

```
verify string c_key (cl_id, cl_detail,
cl_key, cl_tsp);
function check smart contract clchk(
string cl_id, string cl_detail,
string cl_key, string cl_tsp) {
    return cl_id.exist;
    return cl_detail.exist;
    return cl_key.exist;
    return cl_tsp.exist;
}
```

Scheme 10. The smart contracts clins and clchk.

The Licensee calculates:

$$BC_{C-L} = h(r_{C-L}, s_{C-L}), \tag{108}$$

(ID_{BC}, BC_{C-L}) will also be uploaded to the blockchain center. Finally, the APP calculates:

$$P_m = C_2 - C_1^{-b} \tag{109}$$

to successfully obtain the identity of the digital content. This step is performed automatically by the smart contract, and the Licensee cannot skip the verification process privately.

3.6.2. Case 2: Proxy Authorization

After payment, the bank must sign and issue the payment certificate to the Licensee. The Licensee submits the payment certificate to the Proxy, and the Proxy transfers it to the Content Administrator for verification. The CA then authenticates the Licensee's identity and bank payment certificate. After that, the CA generates the authorized key, making time-sensitive tokens. After authorization, the Licensee's application (reader or player) can use the authorized key to automatically decrypt the symmetry key. The APP can browse digital content normally. We present the flowchart of the payment verification and browsing phase (L to P) in Figure 12, the flowchart of the payment verification and browsing phase (P to CA) in Figure 13, the flowchart of the payment verification and browsing phase (CA to P) in Figure 14, and the flowchart of the payment verification and browsing phase (P to L) in Figure 15.

- Step 1: The Licensee generates a random value k_{L-P}, calculates:

$$z_{L-P} = h(ID_L, M_{L-P}, Cert_L, TID, Cert_{pay}, TS_{L-P}, ID_{BC}), \tag{110}$$

$$(x_{L-P}, y_{L-P}) = k_{L-P}G, \tag{111}$$

$$r_{L-P} = x_{L-P} \bmod n, \tag{112}$$

$$s_{L-P} = k_{L-P}^{-1}(z_{L-P} + r_{L-P}d_L) \bmod n, \tag{113}$$

$$Enc_{L-P} = E_{PK_P}(ID_L, M_{L-P}, Cert_L, TID, Cert_{pay}, TS_{L-P}, ID_{BC}), \tag{114}$$

and sends $ID_L, Enc_{L-P}, (r_{L-P}, s_{L-P})$ to the proxy.

- Step 2: The Proxy first calculates:

$$(ID_L, M_{L-P}, Cert_L, TID, Cert_{pay}, TS_{L-P}, ID_{BC}) = D_{SK_P}(Enc_{L-P}), \tag{115}$$

uses

$$TS_{NOW} - TS_{L-P} \leq \Delta T \tag{116}$$

to confirm whether the timestamp is valid, verifies $Cert_L$ with PK_L and $Cert_{pay}$ with PK_{BANK}, verifies the correctness of the ECDSA signature, and then calculates:

$$z_{L-P}' = h(ID_L, M_{L-P}, Cert_L, TID, Cert_{pay}, TS_{L-P}, ID_{BC}), \tag{117}$$

$$u_{L-P1} = z_{L-P}' s_{L-P}^{-1} \bmod n, \tag{118}$$

$$u_{L-P2} = r_{L-P} s_{L-P}^{-1} \bmod n, \tag{119}$$

$$(x_{L-P}', y_{L-P}') = u_{L-P1} G + u_{L-P2} Q_L, \tag{120}$$

$$x_{L-P}' \stackrel{?}{=} r_{L-P} \bmod n. \tag{121}$$

Licensee (L)	Proxy (P)
Choose a random number k_{L-P}	
$z_{L-P} = h(ID_L, M_{L-P}, Cert_L, TID, Cert_{pay}, TS_{L-P}, ID_{BC})$	
$(x_{L-P}, y_{L-P}) = k_{L-P} G$	
$r_{L-P} = x_{L-P} \bmod n$	
$s_{L-P} = k_{L-P}^{-1}(z_{L-P} + r_{L-P} d_L) \bmod n$	
$Enc_{L-P} = E_{PK_P}(ID_L, M_{L-P}, Cert_L, TID, Cert_{pay}, TS_{L-P}, ID_{BC})$	
$ID_L, Enc_{L-P}, (r_{L-P}, s_{L-P})$ $\longrightarrow$	
	$(ID_L, M_{L-P}, Cert_L, TID, Cert_{pay}, TS_{L-P}, ID_{BC}) = D_{SK_P}(Enc_{L-P})$
	Check $TS_{NOW} - TS_{L-P} \leq \Delta T$
	Verify $Cert_L$ with PK_L and $Cert_{pay}$ with PK_{BANK}
	$z_{L-P}' = h(ID_L, M_{L-P}, Cert_L, TID, Cert_{pay}, TS_{L-P}, ID_{BC})$
	$u_{L-P1} = z_{L-P}' s_{L-P}^{-1} \bmod n$
	$u_{L-P2} = r_{L-P} s_{L-P}^{-1} \bmod n$
	$(x_{L-P}', y_{L-P}') = u_{L-P1} G + u_{L-P2} Q_L$
	Check $x_{L-P}' \stackrel{?}{=} r_{L-P} \bmod n$
	If valid, call smart contract lpins and lpchk
	$BC_{L-P} = h(r_{L-P}, s_{L-P})$
	Upload (ID_{BC}, BC_{L-P})

Figure 12. Payment verification and browsing phase (L to P).

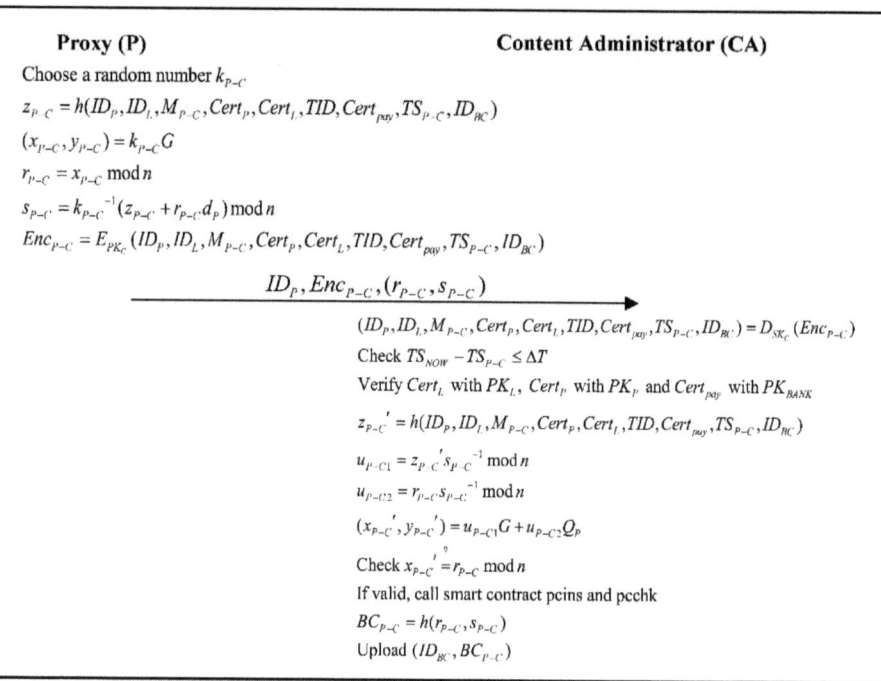

Figure 13. Payment verification and browsing phase (P to CA).

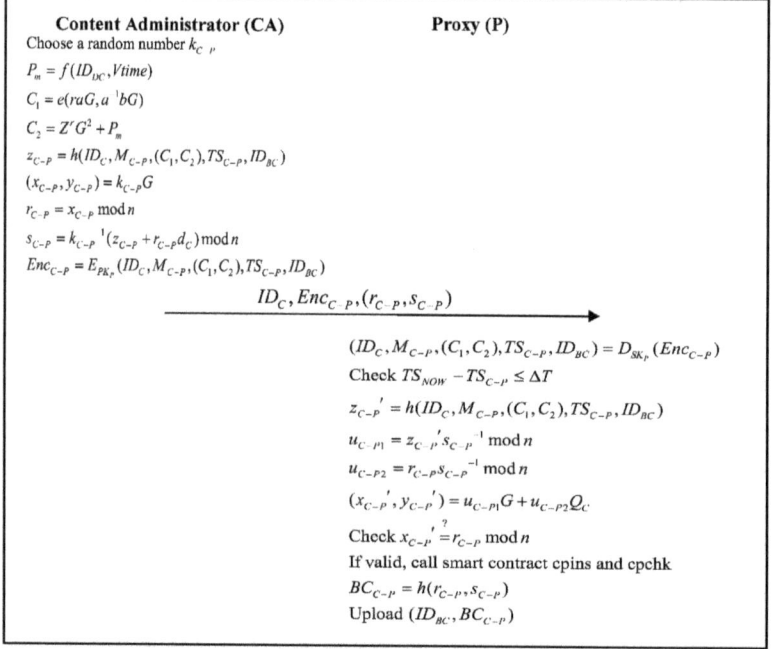

Figure 14. Payment verification and browsing phase (CA to P).

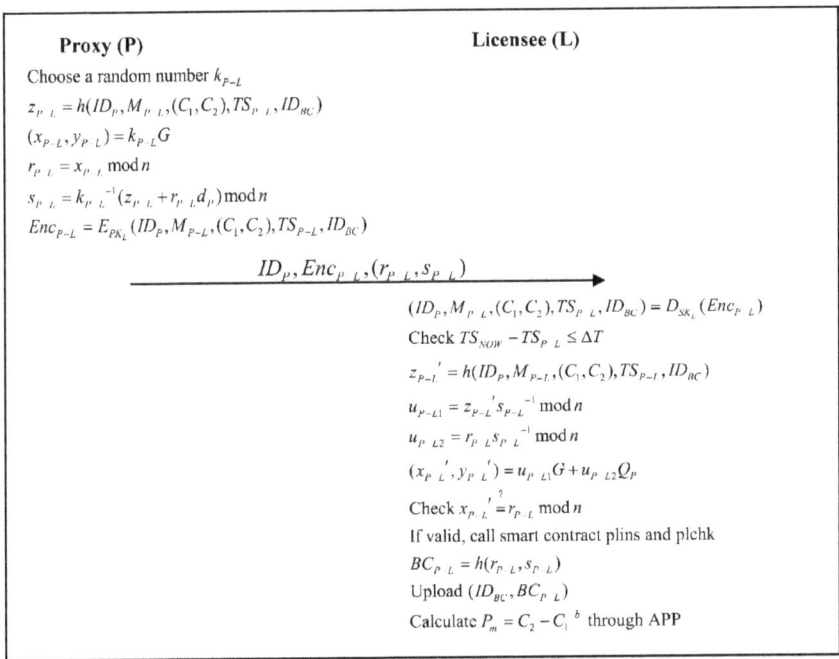

Figure 15. Payment verification and browsing phase (P to L).

If the verification is passed, the proxy will get the relevant payment information and trigger the smart contracts lpins and lpchk. The content is as follows (Scheme 11):

```
function insert smart contract lpins(
string lp_id, string lp_detail,
string lp_payment, string lp_tsp) {
    count ++;
    lp[count].id = id;
    lp[count].detail = detail;
    lp[count].payment = payment;
    lp[count].tsp = tsp;
}
sign string l_key (lp_id, lp_detail,
lp_payment, lp_tsp);
```

```
verify string l_key (lp_id, lp_detail,
lp_payment, lp_tsp);
function check smart contract lpchk(
string lp_id, string lp_detail,
string lp_payment, string lp_tsp) {
    return lp_id.exist;
    return lp_detail.exist;
    return lp_payment.exist;
    return lp_tsp.exist;
}
```

Scheme 11. The smart contracts lpins and lpchk.

The Proxy calculates:

$$BC_{L-P} = h(r_{L-P}, s_{L-P}), \qquad (122)$$

(ID_{BC}, BC_{L-P}) will also be uploaded to the blockchain center.

- Step 3: The Proxy then generates a random value k_{P-C} and calculates:

$$z_{P-C} = h(ID_P, ID_L, M_{P-C}, Cert_P, Cert_L, TID, Cert_{pay}, TS_{P-C}, ID_{BC}), \qquad (123)$$

$$(x_{P-C}, y_{P-C}) = k_{P-C}G, \qquad (124)$$

$$r_{P-C} = x_{P-C} \bmod n, \qquad (125)$$

$$s_{P-C} = k_{P-C}^{-1}(z_{P-C} + r_{P-C}d_P) \bmod n, \tag{126}$$

$$Enc_{P-C} = E_{PK_C}(ID_P, ID_L, M_{P-C}, Cert_P, Cert_L, TID, Cert_{pay}, TS_{P-C}, ID_{BC}), \tag{127}$$

and sends $ID_P, Enc_{P-C}, (r_{P-C}, s_{P-C})$ to the content administrator.

- Step 4: The CA first calculates:

$$(ID_P, ID_L, M_{P-C}, Cert_P, Cert_L, TID, Cert_{pay}, TS_{P-C}, ID_{BC}) = D_{SK_C}(Enc_{P-C}), \tag{128}$$

uses

$$TS_{NOW} - TS_{P-C} \leq \Delta T \tag{129}$$

to confirm whether the timestamp is valid, verifies $Cert_L$ with PK_L, $Cert_P$ with PK_P and $Cert_{pay}$ with PK_{BANK}, verifies the correctness of the ECDSA signature, and then calculates:

$$z_{P-C}' = h(ID_P, ID_L, M_{P-C}, Cert_P, Cert_L, TID, Cert_{pay}, TS_{P-C}, ID_{BC}), \tag{130}$$

$$u_{P-C1} = z_{P-C}' s_{P-C}^{-1} \bmod n, \tag{131}$$

$$u_{P-C2} = r_{P-C} s_{P-C}^{-1} \bmod n, \tag{132}$$

$$(x_{P-C}', y_{P-C}') = u_{P-C1}G + u_{P-C2}Q_P, \tag{133}$$

$$x_{P-C}' \stackrel{?}{=} r_{P-C} \bmod n. \tag{134}$$

If the verification is passed, the content administrator will get the relevant payment information and trigger the smart contracts pcins and pcchk. The content is as follows (Scheme 12):

```
function insert smart contract pcins(
string pc_id, string pc_detail,
string pc_payment, string pc_tsp) {
    count ++;
    pc[count].id = id;
    pc[count].detail = detail;
    pc[count].payment = payment;
    pc[count].tsp = tsp;
}
sign string p_key (pc_id, pc_detail,
pc_cert, pc_tsp);
```

```
verify string p_key (pc_id, pc_detail,
pc_payment, pc_tsp);
function check smart contract pcchk(
string pc_id, string pc_detail,
string pc_payment, string pc_tsp) {
    return pc_id.exist;
    return pc_detail.exist;
    return pc_payment.exist;
    return pc_tsp.exist;
}
```

Scheme 12. The smart contracts pcins and pcchk.

The content administrator calculates:

$$BC_{P-C} = h(r_{P-C}, s_{P-C}), \tag{135}$$

(ID_{BC}, BC_{P-C}) will also be uploaded to the blockchain center.

- Step 5: The content administrator generates a random value k_{C-P} and calculates:

$$P_m = f(ID_{DC}, Vtime), \tag{136}$$

$$C_1 = e(raG, a^{-1}bG), \tag{137}$$

$$C_2 = Z^r G^2 + P_m, \tag{138}$$

$$z_{C-P} = h(ID_C, M_{C-P}, (C_1, C_2), TS_{C-P}, ID_{BC}), \tag{139}$$

$$(x_{C-P}, y_{C-P}) = k_{C-P}G, \tag{140}$$

$$r_{C-P} = x_{C-P} \bmod n, \tag{141}$$

$$s_{C-P} = k_{C-P}^{-1}(z_{C-P} + r_{C-P}d_C) \bmod n, \quad (142)$$

$$Enc_{C-P} = E_{PK_P}(ID_C, M_{C-P}, (C_1, C_2), TS_{C-P}, ID_{BC}), \quad (143)$$

and sends $ID_C, Enc_{C-P}, (r_{C-P}, s_{C-P})$ to the proxy.

- Step 6: The Proxy first calculates:

$$(ID_C, M_{C-P}, (C_1, C_2), TS_{C-P}, ID_{BC}) = D_{SK_P}(Enc_{C-P}), \quad (144)$$

uses

$$TS_{NOW} - TS_{C-P} \le \Delta T \quad (145)$$

to confirm whether the timestamp is valid, it verifies the correctness of the ECDSA signature, and then calculates:

$$z_{C-P}' = h(ID_C, M_{C-P}, (C_1, C_2), TS_{C-P}, ID_{BC}), \quad (146)$$

$$u_{C-P1} = z_{C-P}' s_{C-P}^{-1} \bmod n, \quad (147)$$

$$u_{C-P2} = r_{C-P} s_{C-P}^{-1} \bmod n, \quad (148)$$

$$(x_{C-P}', y_{C-P}') = u_{C-P1}G + u_{C-P2}Q_C, \quad (149)$$

$$x_{C-P}' \stackrel{?}{=} r_{C-P} \bmod n. \quad (150)$$

If the verification is passed, the payment information is confirmed by the content administrator, and the smart contracts cpins and cpchk will be sent. The content is as follows (Scheme 13):

```
function insert smart contract cpins(
string cp_id, string cp_detail,
string cp_key, string cp_tsp) {
    count ++;
    cp[count].id = id;
    cp[count].detail = detail;
    cp[count].key = key;
    cp[count].tsp = tsp;
}
sign string c_key (cp_id, cp_detail,
cp_key, cp_tsp);
```

```
verify string c_key (cp_id, cp_detail,
cp_key, cp_tsp);
function check smart contract cpchk(
string cp_id, string cp_detail,
string cp_key, string cp_tsp) {
    return cp_id.exist;
    return cp_detail.exist;
    return cp_key.exist;
    return cp_tsp.exist;
}
```

Scheme 13. The smart contracts cpins and cpchk.

The Proxy calculates:

$$BC_{C-P} = h(r_{C-P}, s_{C-P}), \quad (151)$$

(ID_{BC}, BC_{C-P}) will also be uploaded to the blockchain center.

- Step 7: The Proxy generates a random value k_{P-L} and calculates:

$$z_{P-L} = h(ID_P, M_{P-L}, (C_1, C_2), TS_{P-L}, ID_{BC}), \quad (152)$$

$$(x_{P-L}, y_{P-L}) = k_{P-L}G, \quad (153)$$

$$r_{P-L} = x_{P-L} \bmod n, \quad (154)$$

$$s_{P-L} = k_{P-L}^{-1}(z_{P-L} + r_{P-L}d_P) \bmod n, \quad (155)$$

$$Enc_{P-L} = E_{PK_L}(ID_P, M_{P-L}, (C_1, C_2), TS_{P-L}, ID_{BC}), \quad (156)$$

and sends $ID_P, Enc_{P-L}, (r_{P-L}, s_{P-L})$ to the Licensee.
- Step 8: The Licensee first calculates:

$$(ID_P, M_{P-L}, (C_1, C_2), TS_{P-L}, ID_{BC}) = D_{SK_L}(Enc_{P-L}), \tag{157}$$

uses

$$TS_{NOW} - TS_{P-L} \leq \Delta T \tag{158}$$

to confirm whether the timestamp is valid, verifies the correctness of the ECDSA signature, and then calculates:

$$z_{P-L}' = h(ID_P, M_{P-L}, (C_1, C_2), TS_{P-L}, ID_{BC}), \tag{159}$$

$$u_{P-L1} = z_{P-L}' s_{P-L}^{-1} \bmod n, \tag{160}$$

$$u_{P-L2} = r_{P-L} s_{P-L}^{-1} \bmod n, \tag{161}$$

$$(x_{P-L}', y_{P-L}') = u_{P-L1} G + u_{P-L2} Q_P, \tag{162}$$

$$x_{P-L}' \stackrel{?}{=} r_{P-L} \bmod n. \tag{163}$$

If the verification is passed, the authorization information is confirmed by Licensee, and the smart contracts plins and plchk will be sent. The content is as follows (Scheme 14):

```
function insert smart contract plins(
string pl_id, string pl_detail,
string pl_tid, string pl_tsp) {
    count ++;
    pl[count].id = id;
    pl[count].detail = detail;
    pl[count].key = key;
    pl[count].tsp = tsp;
}
sign string p_key (pl_id, pl_detail,
pl_key, pl_tsp);
```

```
verify string p_key (pl_id, pl_detail,
pl_key, pl_tsp);
function check smart contract plchk(
string pl_id, string pl_detail,
string pl_key, string pl_tsp) {
    return pl_id.exist;
    return pl_detail.exist;
    return pl_key.exist;
    return pl_tsp.exist;
}
```

Scheme 14. The smart contracts plins and plchk.

The Licensee calculates:

$$BC_{P-L} = h(r_{P-L}, s_{P-L}), \tag{164}$$

(ID_{BC}, BC_{P-L}) will also be uploaded to the blockchain center. Finally, the APP calculates:

$$P_m = C_2 - C_1^{-b} \tag{165}$$

to obtain the identity of the digital content successfully. This step is performed automatically by the smart contract, and the Licensee cannot skip the verification process privately.

4. Analysis

In this section, we analyze the requirements of digital rights management as follows.

4.1. Verifiable

Using digital certificate verification can publicly verify the identity of the Licensee, and the authorization information was published based on the openness and transparency of the information on the chain, truly realizing the high efficiency and specialization in the field of digital copyright.

Let's take the message transmitted by the Licensee (L) and Content Administrator (CA) as an example. When CA sends a message signed by ECDSA to L, L first verifies the correctness of the time stamp and signature, then generates blockchain data $BC_{C-L} = h(r_{C-L}, s_{C-L})$, and uses ID_{BC} as an index to upload the blockchain data to the Blockchain Center (BCC). That is to say, after verifying the correctness of the time stamp and signature for each role that receives the message, it also verifies the correctness of the blockchain data generated by the previous role. Therefore, our proposed solution achieves the characteristics of public verification through blockchain technology and ECDSA digital signature.

4.2. Trustless

The identity of the authorized object of digital content is verified by the Digital Content Administrator. The authorization period is controlled by the Digital Content Administrator. The Licensee cannot occupy or transfer privately. Any nodes that participate in the system do not need to trust each other. The operation of the system and operating rules are open and transparent, and all information is open. A node cannot deceive other nodes. In this way, the trust relationship between nodes is realized, making it possible to obtain trust between nodes at a low cost. For example, when Licensee (L) requests digital content authorization from the Content Administrator (CA), CA will send an authorization message to L. This message $P_m = f(ID_{DC}, Vtime)$ contains the digital content ID and the authorization period, and L will be unable to privately occupy or transfer digital content privately.

4.3. Unforgery

Use time stamp and signature mechanism to irreversibly generate a string composed of random numbers and letters for the data placed in each block. This original text cannot be inferred from the string, thus effectively solving the trust problem. After the hash function operation, the messages are described as follows.

$$z_{L-M} = h(ID_L, M_{L-M}, Cert_L, TS_{L-M}, ID_{BC})$$
$$z_{M-L} = h(ID_M, M_{M-L}, TID, invoice, TS_{M-L}, ID_{BC})$$
$$z_{L-A} = h(ID_L, M_{L-A}, Cert_L, TS_{L-A}, ID_{BC})$$
$$z_{A-M} = h(ID_A, ID_L, M_{A-M}, Cert_A, Cert_L, TS_{A-M}, ID_{BC})$$
$$z_{M-A} = h(ID_M, M_{M-A}, TID, invoice, TS_{M-A}, ID_{BC})$$
$$z_{A-L} = h(ID_A, M_{A-L}, TID, invoice, TS_{A-L}, ID_{BC})$$
$$z_{L-C} = h(ID_L, M_{L-C}, Cert_L, TID, Cert_{pay}, TS_{L-C}, ID_{BC})$$
$$z_{C-L} = h(ID_C, M_{C-L}, (C_1, C_2), TS_{C-L}, ID_{BC})$$
$$z_{L-P} = h(ID_L, M_{L-P}, Cert_L, TID, Cert_{pay}, TS_{L-P}, ID_{BC})$$
$$z_{P-C} = h(ID_P, ID_L, M_{P-C}, Cert_P, Cert_L, TID, Cert_{pay}, TS_{P-C}, ID_{BC})$$
$$z_{C-P} = h(ID_C, M_{C-P}, (C_1, C_2), TS_{C-P}, ID_{BC})$$
$$z_{P-L} = h(ID_P, M_{P-L}, (C_1, C_2), TS_{P-L}, ID_{BC})$$

The hash value cannot be reversed back to the original content, so this agreement achieves the characteristic that the message cannot be tampered with.

4.4. Traceable

After the digital content is on the chain, the data block containing the copyright information is permanently stored on the blockchain and cannot be tampered with. All transaction traces can be traced throughout the entire process, which can be used as a digital certificate to deal with infringement. For example: When we want to verify and trace whether the blockchain data between the Licensee (L) and Content Administrator (CA) is legal, we can compare and verify $BC_{L-C} \stackrel{?}{=} h(r_{L-C}, s_{L-C})$ and $BC_{C-L} \stackrel{?}{=} h(r_{C-L}, s_{C-L})$. When we want to verify and trace whether the blockchain data between the Licensee (L) and Proxy (P) is legal, we can compare and verify $BC_{L-P} \stackrel{?}{=} h(r_{L-P}, s_{L-P})$ and $BC_{P-L} \stackrel{?}{=} h(r_{P-L}, s_{P-L})$. When we want to verify and trace whether the blockchain data between the Proxy (P) and Content Administrator (CA) is legal, we can compare and verify $BC_{P-C} \stackrel{?}{=} h(r_{P-C}, s_{P-C})$ and $BC_{C-P} \stackrel{?}{=} h(r_{C-P}, s_{C-P})$.

4.5. Non-Repudiation

The content of the message sent by each role is signed by the sender with its ECDSA private key. After receiving the message, the receiver will verify the message with the sender's public key. If the message is successfully verified, the sender will not deny the content of the message transmitted. Table 1 is an undeniable description of each role in this program.

Table 1. Non-repudiation of the proposed scheme.

Phase	Signature	Sender	Receiver	Signature Verification
Authentication and issuing invoice phase (direct authorization)	(r_{L-M}, s_{L-M})	L	CA	$x_{L-M}' \stackrel{?}{=} r_{L-M} \bmod n$
	(r_{M-L}, s_{M-L})	CA	L	$x_{M-L}' \stackrel{?}{=} r_{M-L} \bmod n$
Authentication and issuing invoice phase (proxy authorization)	(r_{L-A}, s_{L-A})	L	P	$x_{L-A}' \stackrel{?}{=} r_{L-A} \bmod n$
	(r_{A-M}, s_{A-M})	P	CA	$x_{A-M}' \stackrel{?}{=} r_{A-M} \bmod n$
	(r_{M-A}, s_{M-A})	CA	P	$x_{M-A}' \stackrel{?}{=} r_{M-A} \bmod n$
	(r_{A-L}, s_{A-L})	P	L	$x_{A-L}' \stackrel{?}{=} r_{A-L} \bmod n$
Payment verification and browsing phase (direct authorization)	(r_{L-C}, s_{L-C})	L	CA	$x_{L-C}' \stackrel{?}{=} r_{L-C} \bmod n$
	(r_{C-L}, s_{C-L})	CA	L	$x_{C-L}' \stackrel{?}{=} r_{C-L} \bmod n$
Payment verification and browsing phase (proxy authorization)	(r_{L-P}, s_{L-P})	L	P	$x_{L-P}' \stackrel{?}{=} r_{L-P} \bmod n$
	(r_{P-C}, s_{P-C})	P	CA	$x_{P-C}' \stackrel{?}{=} r_{P-C} \bmod n$
	(r_{C-P}, s_{C-P})	CA	P	$x_{C-P}' \stackrel{?}{=} r_{C-P} \bmod n$
	(r_{P-L}, s_{P-L})	P	L	$x_{P-L}' \stackrel{?}{=} r_{P-L} \bmod n$

4.6. Data Format Standardization

Effectively categorizing digital content and formatting it on the chain helps to effectively manage digital property rights and control the unique authorization power of digital content, and intellectual property rights can be protected. The CA classifies the original multimedia files and encodes them for storage, which will provide fast and consistent authorized content transmission services.

4.7. Timeliness

In our proposed scheme, the Content Administrator (CA) is responsible for the production and management of the digital content property rights and the identity verification of the Licensee (L); the Content Administrator (CA) is also responsible for the issuance of a time-sensitive playback license, and the Licensee's playback key identification code cannot permanently occupy the playback of digital content. The Licensee must obtain the decryption key through the authorization key. However, the authorization key contains the digital content ID and the authorization period. If the authorization period expires, the Licensee will be unable to obtain the decryption key; that is, it cannot perform digital content playback. Thus, we do not worry about the leakage of digital property rights.

4.8. Decentralization/Distribution

In the proposed scheme, the information handled by each role is signed by the role with a private key, and the circulation of all information is open and transparent. A node cannot deceive other nodes. In this way, the trust relationship between nodes is realized, making it possible to obtain trust between nodes at a low cost. Thus, the proposed scheme achieves decentralization and distribution.

4.9. Sustainability

The proposed scheme provides two kinds of authority mechanisms. It not only helps to translate the field visit museum into an online visit to a museum's digital collections, but also promotes social education and contributes to the sustainable operation of the museum via our proposed method.

5. Discussions and Comparisons

5.1. Computation Cost

Table 2 is the computation cost analysis of this scheme.

Table 2. Computation cost analysis of this scheme.

Phase \ Item	BCC	CA	P	L
System role registration phase	$1T_{Mul}$	N/A	N/A	N/A
Authentication and issuing invoice phase (direct authorization)	N/A	$7T_{Mul} + 3T_H$ $+2T_{Cmp} + 2T_{Sig}$	N/A	$7T_{Mul} + 3T_H$ $+1T_{Cmp} + 2T_{Sig}$
Authentication and issuing invoice phase (entrusted authorization)	N/A	$7T_{Mul} + 3T_H$ $+3T_{Cmp} + 2T_{Sig}$	$7T_{Mul} + 3T_H$ $+2T_{Cmp} + 2T_{Sig}$	$7T_{Mul} + 3T_H$ $+1T_{Cmp} + 2T_{Sig}$
Payment verification and browsing phase (direct authorization)	N/A	$9T_{Mul} + 3T_H$ $+3T_{Cmp} + 2T_{Sig}$	N/A	$7T_{Mul} + 3T_H$ $+1T_{Cmp} + 2T_{Sig}$
Payment verification and browsing phase (entrusted authorization)	N/A	$10T_{Mul} + 3T_H$ $+4T_{Cmp} + 2T_{Sig}$	$7T_{Mul} + 3T_H$ $+3T_{Cmp} + 2T_{Sig}$	$7T_{Mul} + 3T_H$ $+1T_{Cmp} + 2T_{Sig}$

T_{Mul}: Multiplication operation; T_H: Hash function operation; T_{Cmp}: Comparison operation; T_{Sig}: Signature operation.

Table 2 is the computation cost analysis of all stages and roles in this scheme. We analyze the payment verification and browsing phase (entrusted authorization) with the highest computational cost. The CA requires 10 multiplication operations, 3 hash function operations, 4 comparison operations, and 2 signature operations. The Proxy requires 7 multiplication operations, 3 hash function operations, 3 comparison operations, and 2 signature operations. The Licensee requires 7 multiplication operations, 3 hash function operations, 1 comparison operation, and 2 signature operations. The method we proposed has a good computational cost.

5.2. Communication Cost

Table 3 analyzes the communication cost of this scheme.

The communication cost analysis of each phase in this scheme is shown in Table 3. We assume that the ECDSA key and signature are 160 bits, the asymmetric message or certificate is 1024 bits, and the rest of the message length such as ID is 80 bits. We analyze the authentication and issuing invoice phase (entrusted authorization) with the highest communication cost. The message sent by the system role to the blockchain center includes 1 other message. The message includes 4 ECDSA keys and signatures, 4 asymmetric messages or certificates, and 4 other messages. The total communication cost in the system role registration phase is 5056 bits, which takes 0.361 ms under 3.5 G (14 Mbps) communication environment, 0.051 ms under 4 G (100 Mbps) communication environment, and takes 0.253 ms under 5 G (20 Mbps) communication environment [34]. The proposed scheme has excellent performance.

Table 3. Communication cost analysis of our scheme.

Phase \ Item	Message Length	Rounds	3.5G (14 Mbps)	4G (100 Mbps)	5G (20 Gbps)
System role registration phase	3552 bits	2	0.254 ms	0.036 ms	0.178 us
Authentication and issuing invoice phase (direct authorization)	2528 bits	2	0.181 ms	0.025 ms	0.126 us
Authentication and issuing invoice phase (proxy authorization)	5056 bits	4	0.361 ms	0.051 ms	0.253 us
Payment verification and browsing phase (direct authorization)	2528 bits	2	0.181 ms	0.025 ms	0.126 us
Payment verification and browsing phase (proxy authorization)	5056 bits	4	0.361 ms	0.051 ms	0.253 us

5.3. Comparison

In this section, we compare the related works which involved the blockchain and smart contract technologies in Table 4.

Table 4. Comparison of the proposed and existing digital right management surveys.

Authors	Year	Objective	1	2	3	4	5	6	7	8
Zhao et al. [17]	2019	Proposed a YODA-based digital watermark management system.	N	Y	Y	Y	N	Y	N	Y
Ma et al. [18]	2018	Proposed efficient and secure authentication, privacy protection, and multi-signature-based conditional traceability approaches.	Y	Y	Y	Y	Y	N	N	N
Vishwa & Hussain [19]	2018	Presented a decentralized data management framework that ensures user data privacy and control.	Y	N	N	Y	Y	N	N	N
Ma et al. [21]	2018	Proposed a blockchain-based DRM platform with high-level credit and security for the Content provider (CP), the Service provider (SP), and customers.	Y	N	Y	N	N	Y	N	N
Lu et al. [22]	2019	Proposed a scheme for digital rights management of design works using blockchain.	Y	Y	N	Y	Y	Y	N	N
Ours	2020	Proposed an authorization of the museum's collections.	Y	Y	Y	Y	Y	Y	Y	Y

1: Blockchain-focused, 2: Comparative analysis with other approaches using tables, 3: Authentication, 4: Verifiable, 5: Unforgeable, 6: Traceable, 7: Data format standardization, 8: Combine cash flow, Y: Yes, and N: No.

6. Conclusions and Future Works

Under the guidance of the "activation and reproduction" public resource thinking based on this research, the use of a "digital authorization" model for museums to provide the information needed by online users and increase financial resources will be a new trend for the sustainable development of museum operations in the future. This research aims at explicating how to use the authorization model that is in line with the actual development of the museum itself and proposes an authorization mechanism based on the blockchain technology related to a museum's digital rights, to realize the economic benefits of the museum collection based on cultural dissemination and education of the public, thereby ensuring the museum's income maximization direction for the perfect development of the current museum-digital authorization model.

This research provides museum exploration based on direct authorization and proxy authorization combined with a cash flow payment verification mechanism. The signature and time stamp mechanism of cryptography is applied to achieve a non-repudiation mechanism (Table 1), which combines blockchain and smart contracts to achieve verifiability, non-tampering, and traceability; digital signatures and digital certificates are used to solve the non-repudiation of the cash flow. Table 2 shows that this method has a good computational cost, while Table 3 shows that the solution we proposed has a low communication cost and can improve the effectiveness of authorization. Table 4 shows the comparison between this digital right management and the existing digital right management survey and proposes a complete presentation of the digital rights of the museum in combination with the financial flow. In addition to the realization of museum social education, the increased benefits of digital rights are conducive to the long-term operation of the museum; the sustainable development of the museum is expected.

In the future, the research will focus on the establishment of a promotion platform for the authorization mechanism of the alliance chain museum of blockchain technology, to achieve a win-win situation of resource sharing and economic benefits. Besides, the world organization has made the world economy globalized and the international market integrated. It is foreseeable that international economic and trade disputes will emerge endlessly. Governments of various countries have added or strengthened arbitration regulations to resolve disputes involving various profits as future digital property management. If there is a dispute, it can be resolved through the mechanism of international legal arbitration. This research provides a good foundation for future research on the authorization of the museum collection alliance chain and the dispute resolution arbitration mechanism.

Author Contributions: The authors' contributions are summarized below. Y.-C.W. and C.-L.C. made substantial contributions to the conception and design. C.-L.C. and Y.-Y.D. were involved in drafting the manuscript. C.-L.C. and Y.-Y.D. acquired data and analysis and conducted the interpretation of the data. The critically important intellectual contents of this manuscript were revised by Y.-C.W. All authors have read and agreed to the published version of the manuscript.

Funding: This research was supported by the Ministry of Science and Technology, Taiwan, R.O.C., under contract number MOST 109-2221-E-324-021.

Informed Consent Statement: This study only base on the theoretical basic research. It is not involving humans.

Data Availability Statement: The data used to support the findings of this study are available from the corresponding author upon request.

Conflicts of Interest: The authors declare no conflict of interest.

Abbreviations

Abbreviations are used in this paper and listed as follows:

q	A k-bit prime number
$GF(q)$	Finite group q
E	The elliptic curve defined on finite group q
G	A generating point based on the elliptic curve E
ID_x	A name representing identity x
k_x	A random value on elliptic curve
(r_x, s_x)	Elliptic curve signature value of x
$M_{x\text{-}y}$	A message from x to y
ID_{BC}	An index value of blockchain message
BC_x	Blockchain message of x
PK_X/SK_X	An asymmetric public/private key
$E_{PKX}(M)$	Use X's public key PKx to encrypt the message M

$D_{SKX}(M)$	Use X's private key SKx to decrypt the message M
TID	The transaction identity
ID_{DC}	An identity of digital content
key_m	Asymmetric key containing KeyID and Seed
$Cert_x$	A digital certificate of x conforms to the X.509 standard
$h(.)$	Hash function
$A \stackrel{?}{=} B$	Verify whether A is equal to B

References

1. Parry, R. *Recoding the Museum: Digital Heritage and the Technologies of Change*; Routledge: London, UK, 2007; pp. 58–81.
2. Fenton, R. *Photographer of the 1850s*; South Bank: London, UK, 1988.
3. The Getty Foundation. Available online: https://www.getty.edu/foundation/initiatives/current/osci/ (accessed on 30 November 2020).
4. Creative Economy Report 2010. United Nations Conference on Trade and Development. Available online: https://unctad.org/system/files/official-document/ditctab20103_en.pdf (accessed on 23 January 2021).
5. Chiou, S.-C.; Wang, Y.-C. The example application of genetic algorithm for the framework of cultural and creative brand design in Tamsui Historical Museum. *Soft Comput.* **2018**, *22*, 2527–2545. [CrossRef]
6. UNESCO. Convention for the Safeguarding of the Intangible Cultural Heritage. 2003. Available online: http://unesdoc.unesco.org/images/0013/001325/132540e.pdf (accessed on 30 November 2020).
7. Chang, C.-W.; Wang, S.-I.; Yang, C.-J.; Shao, K.-T. Fish fauna in subtidal waters adjacent to the National Museum of Marine Biology and Aquarium. *Platax* **2011**, *8*, 41–51. [CrossRef]
8. Liu, M.-C. Image management procedures of the National Museum of Marine Biology and Aquarium. *Museol. Q.* **2013**, *27*. [CrossRef]
9. ARTouch Editorial Department. The Epidemic Is Not Far Away: 1/3 of the US Museums May Be Permanently Closed, and Japanese Exhibitions with No Works. Available online: https://artouch.com/news/content-12951.html (accessed on 26 November 2020).
10. Chen, H.Y.; Wang, H.A.; Lin, C.L. Using watermarks and offline DRM to protect digital images in DIAS. In *Proceedings of the International Conference on Theory and Practice of Digital Libraries*; Springer: Berlin/Heidelberg, Germany, 2007; pp. 529–531.
11. Thomas, T.; Emmanuel, S.; Subramanyam, A.V.; Kankanhalli, M.S. Joint watermarking scheme for multiparty multilevel DRM architecture. *IEEE Trans. Inf. Forensics Secur.* **2009**, *4*, 758–767. [CrossRef]
12. Tsai, M.J.; Luo, Y.F. Service-oriented grid computing system for digital rights management (GC-DRM). *Expert Syst. Appl.* **2009**, *36*, 10708–10726. [CrossRef]
13. Chen, C.L. A secure and traceable E-DRM system based on mobile device. *Expert Syst. Appl.* **2008**, *35*, 878–886. [CrossRef]
14. Chen, C.L. An all-in-one mobile DRM system design. *Int. J. Innov. Comput. Inf. Control* **2010**, *6*, 897–911.
15. Chen, C.L.; Tsaur, W.J.; Chen, Y.Y.; Chang, Y.C. A secure mobile DRM system based on cloud architecture. *Comput. Sci. Inf. Syst.* **2014**, *11*, 925–941. [CrossRef]
16. Hassan, H.E.R.; Tahoun, M.; ElTaweel, G.S. A robust computational DRM framework for protecting multimedia contents using AES and ECC. *Alex. Eng. J.* **2020**, *59*, 1275–1286. [CrossRef]
17. Zhao, B.; Fang, L.; Zhang, H.; Ge, C.; Meng, W.; Liu, L.; Su, C. Y-DWMS: A digital watermark management system based on smart contracts. *Sensors* **2019**, *19*, 3091. [CrossRef]
18. Ma, Z.; Jiang, M.; Gao, H.; Wang, Z. Blockchain for digital rights management. *Future Gener. Comput. Syst.* **2018**, *89*, 746–764. [CrossRef]
19. Vishwa, A.; Hussain, F.K. A blockchain based approach for multimedia privacy protection and provenance. In Proceedings of the 2018 IEEE Symposium Series on Computational Intelligence (SSCI), Bengaluru, India, 18–21 November 2018; pp. 1941–1945.
20. Ma, Z.; Huang, W.; Bi, W.; Gao, H.; Wang, Z. A master-slave blockchain paradigm and application in digital rights management. *China Commun.* **2018**, *15*, 174–188. [CrossRef]
21. Ma, Z.; Huang, W.; Gao, H. Secure DRM scheme based on Blockchain with high credibility. *Chin. J. Electron.* **2018**, *27*, 1025–1036. [CrossRef]
22. Lu, Z.; Shi, Y.; Tao, R.; Zhang, Z. Blockchain for digital rights management of design works. In Proceedings of the 2019 IEEE 10th International Conference on Software Engineering and Service Science (ICSESS), Beijing, China, 18–20 October 2019; pp. 596–603.
23. American Association of Museums. *Museums for a New Century, a Report of the Commission on Museums for a New Century*; American Association of Museums: Washington, DC, USA, 1984.
24. Ma, Z. Digital rights management: Model, technology and application. *China Commun.* **2017**, *14*, 156–167.
25. Du Toit, J. Protecting private data using digital rights management. *J. Inf. Warf.* **2018**, *17*, 64–77.
26. Mrabet, H.; Belguith, S.; Alhomoud, A.; Jemai, A. A survey of IoT security based on a layered architecture of sensing and data analysis. *Sensors* **2020**, *20*, 3625. [CrossRef]
27. Szabo, N. Smart contracts: Building blocks for digital markets. *EXTROPY J. Transhumanist Thought* **1996**, *18*, 16.
28. Szabo, N. The Idea of Smart Contracts. 1997. Available online: http://www.fon.hum.uva.nl/rob/Courses/InformationInSpeech/CDROM/Literature/LOTwinterschool2006/szabo.best.vwh.net/smart_contracts_idea.html (accessed on 26 November 2020).

29. Han, W.; Zhu, Z. An ID-based mutual authentication with key agreement protocol for multiserver environment on elliptic curve cryptosystem. *Int. J. Commun. Syst.* **2014**, *27*, 1173–1185. [CrossRef]
30. Boneh, D.; Lynn, B.; Shacham, H. Short signatures from the Weil pairing. In *Proceedings of the International Conference on the Theory and Application of Cryptology and Information Security*; Springer: Heidelberg/Berlin, Germany, 2001; pp. 514–532.
31. Chen, C.-L.; Yang, T.-T.; Chiang, M.-L.; Shih, T.-F. A privacy authentication scheme based on cloud for medical environment. *J. Med. Syst.* **2014**, *38*, 143. [CrossRef]
32. Chen, C.-L.; Yang, T.-T.; Shih, T.-F. A secure medical data exchange protocol based on cloud environment. *J. Med. Syst.* **2014**, *38*, 112. [CrossRef]
33. Blaze, M.; Bleumer, G.; Strauss, M. Divertible protocols and atomic proxy cryptography. In *Proceedings of the International Conference on the Theory and Applications of Cryptographic Techniques*; Springer: Berlin/Heidelberg, Germany, 1998; pp. 127–144.
34. Marcus, M.J. 5G and IMT for 2020 and beyond. *IEEE Wirel. Commun.* **2015**, *22*, 2–3. [CrossRef]

MDPI
St. Alban-Anlage 66
4052 Basel
Switzerland
Tel. +41 61 683 77 34
Fax +41 61 302 89 18
www.mdpi.com

Applied Sciences Editorial Office
E-mail: applsci@mdpi.com
www.mdpi.com/journal/applsci